Alexander BOTTS

and the EARTHWORM TRACTOR

BOTTS BREAKS HOLLYWOOD

William Hazlett Upson

OCTANE
PRESS

Octane Press, Edition 1.0, October 2021
Volume 3

Main cover illustration by Nick Harris (Beehive Illustration)
Cover portrait by Walter Skor
Interior illustrations by Tony Sarg

ISBN: 978-1-64234-073-0
ePub ISBN: 978-1-64234-074-7
LCCN: 2021936279

Project Edited by Catherine Mandel
Design by Tom Heffron
Copyedited by Maria Edwards
Proofread by Meagan Smith

Austin, TX
octanepress.com

Printed in the United States

CONTENTS

FOREWORD

Alexander Botts on the Silver Screen

BY MARIA EDWARDS

CUE THE LIGHTS AND CAMERAS as Alexander Botts leaps into action to sell those "absolutely sensational" Earthworm tractors on the big screen. Sprung from author William Hazlett Upson's early career with the Caterpillar Tractor Company, Alexander Botts and the Earthworm Tractor were introduced to the world in 1927 via *The Saturday Evening Post* and became a wildly popular series that spawned books, a play, a comic strip, radio adaptations, and the 1936 movie *Earthworm Tractors*, which brought to life Upson's quirky salesman.

Upson was involved with the movie production in the early stages and used this experience to pen the last two Botts yarns in this volume that contains stories dating from 1932–1935. Upson also wrote an essay that details his time in the industry titled "Why Hollywood Drives You Crazy," which appeared in the December 21, 1935, issue of *The Saturday Evening Post*. His enchantment with the hustle and bustle of show business inspired him to give Botts the same starry-eyed passion to see his name in the bright lights. So when Warner Brothers Pictures offered Upson a job to help spruce up the *Earthworm Tractors* screenplay after the original version was rejected by studio executives, he jumped at the opportunity, ignoring the naysayer writers who had been unceremoniously shown the door to Tinseltown.

Upson entered the Hollywood scene not as a writer but as a general adviser and technical expert of tractors. He worked with the supervisor in charge of the film, James Seymour, and chief screenwriter Joel Sayre (author of the *Gunga Din* screenplay), who Upson claims was a brilliant writer but knew nothing about tractors. The trio got along superbly and worked enthusiastically on stringing a series of wisecracks and gags into what they hoped would be an acclaimed slapstick comedy.

As they were finishing up the screenplay, however, the Warner Brothers themselves arrived in person at the studio from New York and essentially claimed the movie was too lowbrow for the American public. They suspended the movie indefinitely and reassigned the lead actor to the studio's production of *A Midsummer Night's Dream*. Upson was left scrambling in Hollywood for a new writing position. After months of his agents failing to secure him a new motion-picture job, Upson and his family headed back East so that he could take up writing for *The Saturday Evening Post* again. Like Botts, and the many other jilted writers before him, Upson was given the cold shoulder by the ever-mercurial Hollywood.

But, as they say in showbiz, the show must go on, and the studio ultimately decided to move forward with the film—without Upson and Sayre. After a whole new team of writers had been hired, *Earthworm Tractors* premiered seven months after Upson's article was published.

The movie portrays Botts at the start of his Earthworm tractor-selling career and utilizes an all-star cast to tell his story. Nineteen thirties film star Joe E. Brown (1891–1973) uses his spot-on comedic timing and amiable screen presence to effortlessly embody the voracious and often bumbling (self-proclaimed) "natural-born salesman" Alexander Botts. Brown threw himself into the role and insisted on doing his own stunts and even learned how to operate the Caterpillar crawlers on the set. Starring opposite Brown is the masterful Guy Kibbee (1882–1956) as Sam Johnson, the blistering, cantankerous customer who is lucky enough to be at the receiving end of all of Botts's sales tricks. Finally, June Travis (1914–2008) rounds out the cast as Mabel Johnson, who guides Botts to success past her father Sam's disdain for all things machine.

While the medium of entertainment might be different, the wild antics remain the same. Botts's very first Earthworm tractor demonstration is a study in disaster as he runs down Johnson's fence, drives the tractor into a swamp, tears off Johnson's truck's radiator while trying to pull it out of said swamp, and subsequently runs clean over the pickup truck with the tractor. What follows once the Earthworm tractor finally comes to a halt is Botts getting a solid knuckle sandwich to the face from one very angry customer. Never fear though, as all natural-born salesmen are wont to do, Botts still manages to make a sale.

Earthworm Tractors blends classic slapstick high jinks, a touch of sweet romance, and some impressive special effects (for the time period) to create an endearing screwball comedy perfect for a lazy weekend afternoon. Contrary to what the Warner Brothers predicted, the picture successfully reached the American public and even premiered in the Peoria, Illinois, theater after being filmed in the Caterpillar plant located in the city.

In addition, the popularity of Botts in *The Saturday Evening Post* drove the audience for this movie, and a favorable review in *Harrison's Reports* labeled it as "a pretty good farce" with "extremely comical" gags that will "provoke howls of laughter." As with Botts on the page, viewers young and old will enjoy watching Botts's Hollywood debut as his adventures play out live on the silver screen.

To carry on the impact Alexander Botts has had on popular culture, Octane Press has acquired William Hazlett Upson's full series of his Alexander Botts stories, some of which did not appear in The Saturday Evening Post. *The collection will presented in its entirety the first time along with original illustrations.*

MORE TROUBLE WITH THE EXPENSE ACCOUNT

ILLUSTRATED BY TONY SARG

EARTHWORM TRACTOR COMPANY
EARTHWORM CITY, ILLINOIS
OFFICE OF THE SALES MANAGER

JANUARY 2, 1932.

MR. ALEXANDER BOTTS,
VICE PRESIDENT IN CHARGE OF TRACTOR SALES,
DEANE SUPPLY COMPANY,
MERCEDILLO, CALIFORNIA.

DEAR BOTTS: Your letter of recent date is received, and we are very glad that you are accepting our offer to reenter the sales department of the Earthworm Tractor Company.

As I have indicated in a former letter, you are to act as a traveling sales-promotion agent. I enclose a list of Earthworm tractor dealers whom we want you to visit, set down in the order in which we wish you to visit them. The first is Mr. George Grubb, Rio Pedro, California. We are writing Mr. Grubb that you will call on him in the near future.

We want you to spend a week or two with each dealer, analyzing his problems, helping him with suggestions and advice, and teaching him the various sales methods which you used in the old days as a salesman for this company, and which you have recently employed so successfully, in spite of the depression, in selling tractors for the Deane Supply Company.

We will pay you four hundred dollars a month, which represents your old salary of five hundred minus the 20 percent cut which we have been forced to make throughout our organization. We will also allow traveling expenses, and in this connection I feel that I should remind you that the good old days of extravagant expense accounts are gone. Our slogan for 1932 is "Net profits are more important than gross volume." We are insisting that all our traveling representatives practice the most rigid economy. I enclose five hundred dollars advance expense money, and suggest that you use it sparingly.

Kindly start as soon as possible, and send us frequent reports of your progress. We have every confidence in you, and wish you the best of luck.

Most sincerely,
GILBERT HENDERSON,
Sales Manager.

ALEXANDER BOTTS
SALES PROMOTION REPRESENTATIVE
EARTHWORM TRACTOR COMPANY

RIO PEDRO, CALIFORNIA.
WEDNESDAY, JANUARY 6, 1932.

MR. GILBERT HENDERSON,
SALES MANAGER,
EARTHWORM TRACTOR COMPANY,
EARTHWORM CITY, ILLINOIS.
VIA AIR MAIL.

DEAR HENDERSON: Well, here I am. I have started in with a rush. You will note that I already have a supply of swell official embossed stationery, which I ordered in advance, and which I am using in this, my first report, to let you know that I arrived in Rio Pedro this morning, that I called on Mr. George Grubb, and that I have already got things moving so satisfactorily that I will need a thousand dollars additional expense money right away. I want you to wire me the money as soon as you receive this letter—otherwise I may be held up in the very important undertakings which I am initiating in this region.

It certainly seems wonderful to be working once more for the good old Earthworm Tractor Company. Of course, it would be even better if I had my wife along, the way I did on the great European trip. But Gadget is pretty busy these days. Alexander Botts, Junior, and Gadget the Second are now almost three years old, and they are the finest pair of twins in the San Joaquin Valley. They are, however, a lot of work, so Mrs. Botts felt she had to stay at home.

But don't get the idea that being alone will cramp my style. I still have just as much brains and ability as ever, and I have a feeling that this new enterprise is going to be the real climax of my career. The fact that I am spending more money than you expected is a most hopeful sign, because it indicates that I am promoting far more activities than you ever dreamed of; although part of the extra expense is due to the fact that our dealer here is a pathetically moribund specimen, who does not know the meaning of the word *cooperate*.

When I introduced myself to Mr. George Grubb this morning, he at once told me that I was simply wasting my time. He said that the tractor business in this region was completely shot, that he hadn't sold a single machine for six months, and that he had no sales in prospect.

"And if I can't sell tractors around here," he said, "nobody can. So you might as well take the next train out of town. I don't want to waste my time listening to a lot of ignorant suggestions from an outsider like you who doesn't understand the conditions here, and probably doesn't know anything about selling tractors anyway."

These ungracious remarks, naturally, pleased me very much. I saw at once that in working with Mr. Grubb I would run into various difficulties. And difficulties always stimulate me, because I can look forward with so much pleasure to the warm glow of satisfaction which is sure to envelop me when I have overcome them.

I proceeded to handle Mr. Grubb with all my old-time tact and adroitness. An ordinary sales representative might have been discouraged at the old guy's opening remarks, or he might have lost his temper and told him—truthfully enough—that he was an egregious ass to turn down in this loutish way a perfectly friendly and well-meant offer of assistance. But I made neither of these mistakes.

"Mr. Grubb," I said, with a pleasant smile, "your remarks are partially true. As yet, I know nothing of conditions in your territory. But I can learn. Accordingly, I plan to take a room at the hotel, stay around a few days and see if I can stir up something of interest. I might even locate a prospect for you. But whatever happens, I will not call on you or bother you in any way until I have something definite to tell you."

"You'd better not," said Mr. Grubb. "And I warn you again that you are just wasting your time. There isn't anybody around here that wants to buy an Earthworm tractor. And even if they did, they haven't got the money to pay for it. There just is no business any more at all."

"Well, I guess I'll look around, anyway," I said. "Good morning, Mr. Grubb."

"Goodbye," he said.

Leaving this poor idiot sitting gloomily in his office, I walked briskly downtown and began nosing about asking questions. Everybody was most pessimistic. Having noticed that business was slightly sick, they had decided it was on its deathbed. Times were terrible, they told me, and nothing was going on at all. Even the new post office, which the Government was just starting to build, was held up because the contractor had been unable to get the proper kind of stone.

When I heard this last bit of information I began to prick up my ears. My subtle intuition told me that here there might possibly be some business for tractors. And I was right. Before long I located a rather sad and disagreeable old party by the name of Ira Button who owns a sandstone quarry about ten miles away up in the mountains, and who had contracted, for a very pleasing sum of money, to supply the stone for the monumental new post office. He had worked the quarry for several months, taking the stone out and getting it ready. Then, last week, just as he was going to start shipping it to town by motor trucks, the highway bridge over the Rio Pedro Canyon was washed out and the only road from his quarry was completely cut off. He said the state highway department would probably build a new bridge, but it would take many months to complete it. And, as the stone had to be delivered within four weeks, it looked as if poor old Mr. Ira Button was completely blown up. So there he sat in his office, drawing little pictures on the blotter, thinking about his troubles and doing nothing at all to remedy them.

"What sort of a place is that canyon?" I asked. "Are the sides straight up and down or sloping?"

"They're sloping," said Mr. Button. "But they're a pretty steep slope."

"Is it possible for a man to climb down into this canyon and up the other side?" I asked.

"Yes."

"Very good," I said. "What you need is an Earthworm tractor. With one of these mechanical marvels you can pull loads over the roughest country. You can put your stone in a wagon, take it down into the canyon, ford the stream and drag it up the other side."

"I thought of that," said Mr. Button, "and I asked Mr. Grubb, the Earthworm tractor dealer, if it could be done."

"What did he say?"

"He said I was crazy. He said the tractor would get smashed to pieces the first trip. And he didn't want to do business with me anyway, because I have no money to pay for a tractor. I've sunk everything I own in the quarry."

"That's too bad," I said. "But you'll have plenty of money, won't you, when you deliver that stone?"

"Yes, if I deliver it."

"Did old Grubb go out and look over the ground with you?"

"No, he said he didn't have time to bother with it."

"All right," I said, "you and I are going to inspect that canyon, Mr. Button. Is that your flivver out in front?"

"Yes."

"Very good. Let's go."

We went. A drive of about ten miles brought us to the edge of the Rio Pedro Canyon—two hundred feet deep, a quarter of a mile wide, with a rapid stream in the bottom, rushing past the wreckage of the old highway bridge.

"The quarry," said Mr. Button, "is right over there on the other side, less than half a mile away. So near, and yet so far."

"All right," I said. "Let's investigate."

After spending several hours scrambling around over the rocks, I located what looked like a perfectly practicable tractor route, and made up my mind to put on a demonstration.

The banks near the old bridge are too steep even for an Earthworm, but by driving about a mile upstream we can make a crossing just above the great Rio Pedro Falls, where the canyon is not more than a hundred feet deep, and where the banks don't have much more than a forty-five-degree slope. After crossing the canyon we will bring the stone to the end of the present road, and it will then be taken to town in motor trucks.

While we were tramping around we met another man who was looking over the canyon. He was particularly interested in the waterfall, which is about a hundred feet high, and very beautiful. He said he was coming up next Saturday with a group of people who are going to take some moving pictures of the place.

This at once gave me another splendid idea. I told the gentleman I expected to be up there myself in a few days with a tractor, and I said I would pay him any reasonable price up to five hundred dollars for a series of shots showing the tractor and the wagon negotiating this difficult and rocky country. The moving-picture man refused to commit himself in advance, but I feel certain he will fall in with my plans.

And my demonstration will thus kill two birds with one stone. It will sell a tractor to Mr. Button, and it will provide our advertising department with a remarkable picture which it can send all over the country to show the skeptics exactly what an Earthworm tractor is capable of when handled by a man who really understands it.

Upon our return to town, I discovered that Mr. Ira Button has no wagon adequate for the rough work I am planning, so I ordered one to be specially made by the local blacksmith and wheelwright. It will be of unusual size and strength, but the cost will be only three hundred dollars, and the maker has promised to rush the work and have it ready by Saturday morning.

After arranging for the wagon, I called on Mr. George Grubb, our dealer, and laid the whole beautiful scheme before him, suggesting that he lend me one of his tractors and stand the expense of the wagon. But, as I had feared, he failed to become enthusiastic.

"The country around that canyon is entirely too rough for tractors, Earthworms or any other kind," he said. "And I certainly won't let you use one of my stock machines to put on any such demonstration. You would just knock the thing all to pieces. And I have no intention of paying for that fool wagon you say you have ordered. The whole plan is idiotic. This man Button can't buy a tractor. He hasn't got the money to pay for it."

"He'll have the money," I said, "as soon as he delivers that stone."

At this Mr. Button laughed in what I can only describe as a jeering manner. "If he had the faintest chance of getting that stone down here," he said, "I might trust him. But he hasn't. So, if he wants to buy a tractor from me, he'll have to pay cash—which he can't. So that's the end of it."

"Not at all," I replied. "I have been sent to help you, and I am going to help you in spite of yourself. You have some Earthworm tractors on hand here in your warehouse?"

"I have just one—a sixty-horsepower model."

"Exactly what I want," I said. "Here is my offer: I will rent this machine from you for one or more days, starting next Saturday, at twenty-five dollars a day. I will pay for the wagon I have ordered. I will haul out enough stone for Mr. Button so that he can get a first payment from the people who are building the post office. This will enable him to offer you a first payment on the tractor, and when you see how things are going, I am sure you will be only too glad to sell him the machine. In the remote contingency that he doesn't buy the tractor, the Earthworm Tractor Company agrees—through me, its representative—to return the tractor to you in perfect condition or pay you for it in full. Nothing could be more generous than that. You can't lose. What do you say? Will you rent me the tractor?"

"Well," he said, "if you want to make a fool of yourself, and if the Earthworm Tractor Company is willing to pay for your foolishness, it is all right with me."

"Very good, Mr. Grubb," I said. "I'll call for the tractor on Saturday. Good afternoon."

As it was then almost six o'clock, I came back to the hotel and ate supper. I have been spending the evening writing this report. Tomorrow and Friday I will superintend the construction of the wagon and do some more general investigating. And on Saturday the real excitement will start.

I have given you a very full account of my activities, so that you may see that my extra expenses are really a form of enlightened economy. I may not have to spend all of the thousand dollars, but I am having you send it to me anyway, just to be on the safe side. And even if I do use it all, the expenditure of these few paltry dollars will be completely overbalanced by the benefits derived from the moving picture and from the revival of the tractor business which my demonstration will accomplish.

Most sincerely,
ALEXANDER BOTTS.

———

GEORGE GRUBB
EARTHWORM TRACTOR DEALER

RIO PEDRO, CALIFORNIA.
JANUARY 6, 1932.

MR. GILBERT HENDERSON,
SALES MANAGER,
EARTHWORM TRACTOR COMPANY,
EARTHWORM CITY, ILLINOIS.
VIA AIR MAIL.

DEAR MR. HENDERSON: Your representative, Mr. Alexander Botts, has arrived. And as long as you have sent out this man to offer me unsolicited and gratuitous advice on how to run my own business, I am going to take it upon myself to give you a little advice on how to run yours.

Of all the crazy ideas ever evolved by the Earthworm Tractor Company—and there have been many—this latest scheme is the most cockeyed. It wouldn't be so bad if times were good. But you pick out the very moment when business is at its worst and I am already worried to death, and you proceed to annoy me further by sending a lunatic who wants to make a disgusting exhibition of himself by performing a lot of monkeyshines all over my territory.

Mr. Botts blew in here this morning, and already has arranged to give a demonstration hauling stone over a section of mountain country which is so rough that he is sure to smash up the tractor before he can accomplish anything at all. And the purpose of this demonstration is to sell a tractor to a man who has absolutely no money to pay for it. Mr. Botts further tells me that he is going to hire a man to take moving pictures of his demonstration for advertising purposes. A sweet advertisement that will be—a picture showing an Earthworm tractor attempting to drive over a lot of rocks and boulders, and knocking itself to pieces in the attempt.

I always try to be fair with you people, and I am cooperating with Mr. Botts as much as I reasonably can. Upon his assurance that the Earthworm Company would pay all expenses and make good any damages that might ensue, I have told him that he can use for his demonstration a sixty-horsepower Earthworm tractor which I have in stock here.

But I hate to do it. I hate to risk a four-thousand-dollar machine in this way, even if I don't have to stand the loss myself.

And I'm afraid the whole procedure will make the Earthworm tractor ridiculous in the eyes of the public. It is bad business. Instead of wasting your money on wild schemes like this, you might better reduce the price of the tractor and give your hard-working dealers a bigger discount. Think it over.

Very truly,
GEORGE GRUBB.

———

TELEGRAM
EARTHWORM CITY ILL JAN 9 1932
GEORGE GRUBB
RIO PEDRO CAL

AM WIRING BOTTS TO CANCEL THE DEMONSTRATION OF WHICH YOU DISAPPROVE AND TO CONFINE HIS ACTIVITIES TO GIVING YOU SUCH ADVICE AND SUGGESTIONS AS YOU DESIRE STOP OUR ONE THOUGHT IN SENDING MR BOTTS WAS THAT HE MIGHT HELP YOU AND WE STILL HOPE AND BELIEVE THAT HE MAY BE ABLE TO DO SO

GILBERT HENDERSON

TELEGRAM
EARTHWORM CITY ILL JAN 9 1932
ALEXANDER BOTTS
CARE GEORGE GRUBB
RIO PEDRO CAL

YOUR LETTER RECEIVED BUT AM NOT SENDING THE
FUNDS YOU REQUEST STOP REFER TO MY LETTER OF
JANUARY SECOND NET PROFITS RATHER THAN GROSS
SALES THE WATCHWORD FOR NINETEEN THIRTY
TWO STOP EXPENSES MUST BE KEPT DOWN STOP YOU
HAVE NO AUTHORITY TO FINANCE DEMONSTRATIONS
STOP THAT IS THE DEALERS BUSINESS STOP YOU WILL
AT ONCE CANCEL PLANS FOR THE DEMONSTRATION
DESCRIBED IN YOUR LETTER UNLESS GRUBB WILL STAND
ALL EXPENSE STOP YOUR JOB IS TO ASSIST DEALER WITH
ADVICE AND TACTFUL SUGGESTIONS STOP LETTER FROM
GRUBB INDICATES YOU HAVE FAILED TO USE TACT AND
HAVE GIVEN NO SUGGESTIONS WHICH HE CONSIDERS
WORTH ACTING UPON STOP WAKE UP AND TRY TO MAKE
YOURSELF A HELP RATHER THAN A HINDRANCE TO MR
GRUBB STOP WATCH THAT EXPENSE ACCOUNT

GILBERT HENDERSON

———

TELEGRAM
RIO PEDRO CAL JAN 9 1932
GILBERT HENDERSON
SALES MANAGER
EARTHWORM TRACTOR CO
EARTHWORM CITY ILL

YOUR WIRE RECEIVED STOP AM MUCH DISAPPOINTED
IN YOUR ATTITUDE STOP HOW DO YOU EXPECT ME TO
ACCOMPLISH ANYTHING WHEN YOU WONT BACK ME
UP WITH THE NECESSARY FUNDS QUESTION MARK AND
HOW DO YOU EXPECT ME TO HELP THIS GUY WITH

ADVICE AND SUGGESTIONS WHEN HE WONT LISTEN TO ANYTHING I SAY ANOTHER QUESTION MARK AM GOING AHEAD WITH DEMONSTRATION ANYWAY AND WILL SEND YOU FULL REPORT AS SOON AS I HAVE TIME STOP IN HASTE

ALEXANDER BOTTS

———

GEORGE GRUBB
EARTHWORM TRACTOR DEALER

RIO PEDRO, CALIFORNIA.
SATURDAY EVENING, JANUARY 9, 1932.

MR. GILBERT HENDERSON,
SALES MANAGER,
EARTHWORM TRACTOR COMPANY,
EARTHWORM CITY, ILLINOIS.

DEAR MR. HENDERSON: I enclose my bill for $3,521.64, which you owe me for one sixty-horsepower Earthworm tractor. The amount represents the list price of $4,000, plus freight, and minus my dealer's commission.

Things have turned out exactly as I told you they would—only worse. I had expected this wild representative of yours to damage my tractor to some extent. But I had not supposed he would be able to convert it into a total loss.

In case Mr. Botts has not written you of his exploits this afternoon, or in case he has attempted to excuse himself by sending you some highly adorned cock-and-bull story, I will give you a brief account of the facts as I observed them.

Owing to various outside matters of business, I did not reach my office today until two o'clock in the afternoon. When I arrived, my secretary handed me the telegram which you had sent in answer to my letter. My secretary also stated that there had been a telegram for Mr. Botts, which she had given him when he came into the office toward the end of the morning.

"Did Mr. Botts," I asked, "make any remarks after he read his telegram?"

"Yes," she replied. "He said the message gave him a big laugh. He said that his boss at the factory didn't want him to put on this demonstration, but he was going to anyway. He said he had already agreed to buy a special wagon from the blacksmith, and he had promised Mr. Button to haul some stone for him, and he had arranged with a moving-picture man for the taking of a picture. He told me he just didn't have the heart to disappoint all these people, so he took the tractor and drove away."

Upon hearing this, I was, naturally, much displeased. I didn't want Mr. Botts using my tractor to put on a demonstration against your orders. So I at once got into my car and drove out to the Rio Pedro Canyon road with the intention of stopping him. Unfortunately, I was too late.

When I reached the broken highway bridge, I stopped the car. Somewhere in the distance I heard the roar of a tractor motor. As the noise seemed to come from somewhere in the canyon, I climbed down the steep bank to the edge of the river. Then, following the direction of the noise, I walked upstream for almost a mile. Finally I rounded a bend and came in sight of the great Rio Pedro Falls. On the bank of the stream at the top of the falls, I observed Mr. Alexander Botts and my Earthworm tractor. He was evidently preparing to drive across the stream at the very brink of the cataract.

Two motion-picture cameras, with their operators, were perched in advantageous positions on the rocky canyon wall. There appeared to be a number of people present, but I could not tell how many, as my view was limited by the fact that I was looking up from the bottom of the falls.

At once I shouted to Mr. Botts to stop, but the roar of the motor and the thundering of the waters completely drowned out my voice. I waved frantically, but no one noticed me. The walls of the canyon for some distance below the falls are sheer rock. In order to reach the top of the falls I would have had to follow the stream for almost a mile down to the neighborhood of the ruined bridge, then climb out of the canyon and return along the rim. There was no time for this.

As I watched, I saw Mr. Botts step down from the tractor, and, with the assistance of another man, place a very lifelike-looking dummy in the driver's seat. This procedure certainly showed Mr. Botts in his true colors. He was perfectly willing to risk my tractor for the sake of his half-witted advertising motion picture, but he was not willing to risk his own worthless neck. Once more I yelled "Stop" at the top of my voice. But no one heard me. I saw Mr. Botts throw in the clutch. Slowly and steadily, with no one to guide it but the dummy driver, that tractor started moving straight across the rushing stream not more than three feet from the edge of the waterfall. The men with the moving-picture cameras seemed to be working feverishly.

At first, I thought the machine would get across all right. It held its direction straight enough, but it had been aimed a little bit wrong, so that its course, instead of carrying it parallel to the edge, brought it gradually nearer and nearer. Finally, about two-thirds of the way across, it toppled over, and with a sickening plunge disappeared into the depths of the pool at the bottom.

As the disaster was so complete, and as there was, obviously, nothing more I could do, I retraced my steps, climbed out of the canyon and drove back to town.

Mr. Botts has not yet dared to show up here at the office. If he does, I shall simply have him thrown out at once. And in the future I must insist that all business between myself and the Earthworm Tractor Company be handled direct, and not through any such outlandish emissaries as this Botts person.

In conclusion, I draw your attention once more to the enclosed bill. I shall expect your check in full payment by return mail.

Very truly,
GEORGE GRUBB.

———

EARTHWORM TRACTOR COMPANY
EARTHWORM CITY, ILLINOIS
OFFICE OF THE SALES MANAGER

MONDAY, JANUARY 11, 1932.

MR. ALEXANDER BOTTS,
CARE MR. GEORGE GRUBB,
RIO PEDRO, CALIFORNIA.
VIA AIR MAIL.

DEAR BOTTS: Your telegram, stating that you were about to put on a demonstration in spite of my direct orders to the contrary, arrived on Saturday.

Today I have received a letter from Mr. George Grubb, telling of the disastrous consequences of that demonstration, and stating that he expects the Earthworm Company to pay for the tractor which was destroyed in the course of your highly reckless activities.

As we have not yet heard from you, we do not know exactly how many more heavy expenses you have incurred. It is obvious, however, that you have deliberately and completely disregarded my instructions to the effect that your expense account must be kept down to a minimum, and have completely failed to accomplish anything in the way of assisting our dealer, Mr. George Grubb.

In these circumstances, it is necessary for us to advise you that your services will no longer be required by this company. If you care to submit an itemized statement of your expenses, the same will be considered by our accounting department.

> Very truly,
> GILBERT HENDERSON,
> *Sales Manager.*

———

ALEXANDER BOTTS
SALES PROMOTION REPRESENTATIVE
EARTHWORM TRACTOR COMPANY

RIO PEDRO, CALIFORNIA.
WEDNESDAY, JANUARY 13, 1932.

MR. GILBERT HENDERSON,
SALES MANAGER,
EARTHWORM TRACTOR COMPANY,
EARTHWORM CITY, ILLINOIS.
VIA AIR MAIL.

DEAR HENDERSON: I meant to write you before, but I have been very busy. And it is just as well I waited, because now I can answer your snappy little letter, which has just arrived. It certainly seems like the good old days to have you bawling me out so thoroughly and so completely. I sure got to hand it to you, Mr. Henderson. You may be getting old, but you're not losing any of your vitality or any of your command of the English language. When I got to the end of the letter, and saw that you were actually pulling the old bluff of pretending to fire me, it just made me feel good all over.

It also seemed just like the dear old days to have you giving me definite instructions as to what I should do, in spite of the fact that you are so far away that you, naturally, are completely ignorant of conditions here. And I see you are still making the same old error of concentrating your attention too strongly on the size of the expense account. "Net profits rather than gross sales" is a good slogan, but you should always remember that increasing the profits may be accomplished just as well by adding to the receipts as by cutting down the expenses.

In my recent activities here in Rio Pedro, it is true that I ran up a good many expenses. But these expenses were necessary in order that we might receive even greater advantages. When I wrote you my former letter, I had everything organized on a perfectly sound basis. I had every reason to expect that, at an outlay of a few paltry hundreds of dollars, I would sell a tractor, and, by so doing, teach the opinionated and obstinate Mr. Grubb a much-needed lesson in salesmanship, and inspire him to go ahead and make further sales on his own account. In addition, I expected to get a magnificent motion picture for the use of the advertising department. Either one of these achievements would have justified the expense. So, even if no other factors had been involved, I should have probably disregarded your telegram and gone ahead with my demonstration.

But by the time that telegram arrived, I had discovered and taken advantage of certain new factors which were so favorable to my enterprise that it would have been idiotic to hold back. The discovery and utilization of these new factors was due entirely to my own energy and resourcefulness. Early last Saturday morning the blacksmith informed me that the big three-hundred-dollar wagon would be completed a little before noon. So, instead of loafing around town for several hours until it should be ready, I persuaded Mr. Button to drive me out to the canyon for a final inspection of the ground. Incidentally, we had heard that the moving-picture people were already there, and I wanted to talk to them and make definite arrangements for our picture.

When we arrived at the canyon, we discovered a very curious situation. The whole landscape was swarming with humanity. Instead of a small group taking shots of the scenery, as I had expected, there was a large outfit from one of the big studios in Hollywood, engaged in the production of a stupendous and spectacular drama of the great open spaces. Besides the producing and executive staff, there were a lot of stars and featured players, and an appalling mob of extras, including several dozen cowboys and at least five hundred Indians.

Leaving Mr. Button sitting in the car, I began walking around looking for the advance agent I had talked with a few days previously. But I could not find him. Either he was not there or he was lost in this great crowd. I therefore approached a gentleman who was standing beside a camera and told him I wished to have a few private movies taken. He let out a loud laugh. "You'll have to see the director about that," he said, "and from the way the old guy is carrying on this morning, there isn't one chance in a thousand he'll do anything for anybody. Boy, that lad is sore."

"Has there been some trouble?" I asked.

"I'll say there has. According to the plot of this picture, the hero and heroine escape across the canyon. The villain tries to follow them in a big tractor, and he gets swept over the falls. That's the big scene, and it looks now as if it has gone haywire."

"It has?"

"Yes. The director didn't want to smash up a new tractor by sending it over the falls, so he brought along an old secondhand piece of junk. And now this old tractor has broken down completely, and they can't even drive it up to the top of the falls, let alone run it over."

"Ah, ha!" I said at once. "This is a lucky break for me."

It was, indeed, a most unusual and unexpected piece of good fortune. But I wish to point out, for the benefit of certain people in the Earthworm Tractor Company who may be inclined to think that my success as a salesman is due entirely to what they call fool luck, that I never would have discovered this situation unless I had been up on my toes and chasing about the country with my eyes wide open and my mind on the alert. And I never would have been able to exploit the opportunity to its fullest extent, had I not had the skill and finesse to nurse the situation along until the time was ripe to take definite action.

Most people would have rushed to the director at once, and tried to sell him a tractor. But I decided to make a careful and cautious approach. My first move was to walk over to where the director was standing and listen unobtrusively to what was going on.

The director was a large man with a red face, and he seemed in a state of great agitation. "This is a mess," he said. "Just look at all these people standing around, drawing their salaries and doing nothing. We've got to finish up here today, and we can't do it unless we get a tractor."

Various assistants began fluttering about and explaining that it would take two days to repair the tractor, and just as long to get another one up from Los Angeles.

"There must be tractors somewhere in this godforsaken country," he said. "And it's up to some of you guys to locate one."

"If you are looking for a tractor," I said, stepping forward, "I have one down at Rio Pedro. If you want to use it, I can bring it up here."

"How long will it take?" he asked.

"About two hours," I said.

"All right," he said. "How much do you want for it?"

"We'll discuss that when I arrive. Goodbye."

I hurried back to the road and had Mr. Button drive me to town as fast as possible. When I reached Mr. Grubb's office, his secretary handed me your telegram directing me to cancel the demonstration. Naturally, I paid no attention to these instructions, and, after sending you a brief wire in reply, I took the tractor, hooked onto the big wagon which the blacksmith had just finished, and drove as rapidly as possible to the canyon, arriving about three hours after I had left.

During this time, the director—just as I expected—had worked himself up into a state of far greater anxiety and impatience than ever.

Besides, I now had the tractor on hand where he could look at it and realize that it was exactly the machine he needed. All this made it possible for me to bargain with him more successfully than when I first talked to him.

In fact, I now had him where I could make him eat right out of my hand. During my absence he had sent several of his assistants down to scour the farming country around Rio Pedro, and report whether they could find a tractor. Fortunately, he had as yet heard nothing from these people. He had also tried to make the five hundred Indians drag the old broken-down tractor by main strength up to the top of the falls. But after an hour's work, they had moved it only a hundred yards, and gave it up as a bad job.

I was now his only hope, and he was willing to buy the tractor at two or three times the list price. I told him, however, that I did not care to sell, and I proposed an arrangement that I had thought out very carefully. First of all, I insisted that he let me haul one load of stone from the quarry around to the road, in order to show Mr. Button, the quarry man, that it could be done. I also insisted that moving pictures be taken of this trip. At first, the director kicked like a steer, because this would delay everything for about an hour. But I remained firm, and at last he agreed. The trip was entirely successful. We got several thousand feet of wonderful pictures, and the moving-picture company will send you the films, free of charge, as soon as they are developed.

After the trip with the stone was over, I took the tractor up the canyon again. And in consideration of a payment of six thousand dollars in cash, I ran the old baby over the big waterfall. Perhaps I was a fool to do it so cheap. The poor director would have willingly paid more. However, I am not the man to take a mean advantage of someone else's misfortunes, and I am pleased to report that the director thanked me warmly, and stated that he was entirely satisfied with the deal.

I am even more pleased to report an additional achievement. After the moving-picture people departed, I went down to the foot of the waterfall to see what was left of the tractor, and I was delighted to discover that instead of descending upon a heap of boulders, as at Niagara, the waters of this cataract plunge directly into a very deep pool. This at once gave me an idea. And, by working all day Sunday with several of Mr. Grubb's mechanics and a long cable and a winch from Mr. Button's quarry, we were able to hook onto the tractor and drag it up on the bank. A thorough examination showed, just as I had expected, that it takes more than a drop of one hundred feet into a pool of water to destroy a sturdy Earthworm tractor. The radiator, part of the hood, and various minor widgets were smashed up. But a couple of days' work and a few hundred dollars' worth of parts purchased from Mr. Grubb have sufficed to put the machine back in running order, apparently just as good as new.

So everything has worked out even better than I expected. I have paid old man Grubb the full list price of four thousand dollars for his tractor, plus freight, so you don't have to bother about the bill which he sent you. And I have sold the machine and the big wagon to Mr. Button, who will use them to haul out his stone. I have chosen to regard the tractor as second-hand, placing the price, including the wagon, at three thousand dollars cash. This money was advanced by the contractor who is building the post office, as soon as he found out that the tractor was actually able to bring out the stone which he needed.

Mr. Grubb seems to be the most surprised man in all Southern California. The poor sap actually didn't know, until I showed him, that an Earthworm tractor is built strong enough to negotiate this rough mountain country around here. He tells me that as soon as he can get in another machine, he will put on a demonstration for a man in the western part of the county who is engaged in a big logging operation and will probably take several machines as soon as he is shown how well they get over the hills and rocks.

Having disregarded your telegram instructing me to cancel my demonstration, I will now disregard your letter telling me I am fired, and proceed to the next job, with the feeling that I have made a reasonably auspicious start. And I think you ought to agree with me. Possibly you will not have the imagination or the artistic sensibilities necessary to appreciate the remarkable movie I have made. Possibly you may not be much impressed by the sale of the tractor to Mr. Button or by the way I have waked up Mr. Grubb. But certainly, in view of the interminable manner in which you have been harping upon the subject, you cannot fail to appreciate my expense account, which may be summed up as follows:

Received from moving-picture director......................................$6,000.00
Received from Ira Button for tractor and wagon. 3,000.00
Total Receipts. $9,000.00
Paid to Mr. George Grubb, for tractor, gas, oil,
    spare parts, labor, and so on. $4,571.20
Paid to blacksmith, for wagon..300.00

Paid to myself (traveling and miscellaneous expense)......................246.37
Paid to myself, for this month's salary (I thought
I might as well hold this out, so as to save you the
trouble of sending me a check later. You will note
that I have decided to raise my pay to an amount
more nearly corresponding to what I am worth) 500.00
Total Expenses $5,617.66
Total Receipts $9,000.00
Total Expenses 5,617.66
Net profit (which you have been talking about so continually)..... $3,382.34

I enclose a draft for this last amount, and in conclusion I wish to suggest that if you would only hire a few more men like me, you wouldn't have to worry about whether you sold any tractors or not. You could run the whole company on the profits from the salesmen's expense accounts.

Very truly,
ALEXANDER BOTTS.

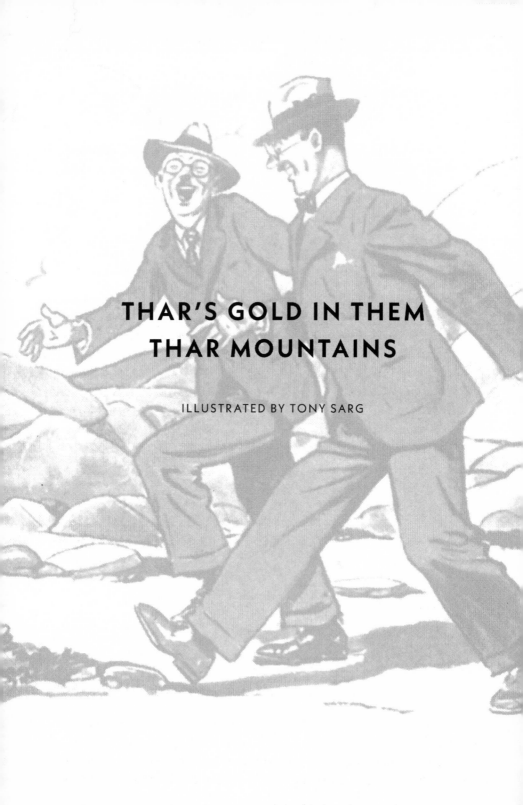

THAR'S GOLD IN THEM THAR MOUNTAINS

ILLUSTRATED BY TONY SARG

ALEXANDER BOTTS
SALES PROMOTION REPRESENTATIVE
EARTHWORM TRACTOR COMPANY

Dow's Gulf, Vermont.
Monday Evening, June 6, 1932.

Mr. Gilbert Henderson,
Sales Manager,
Earthworm Tractor Company,
Earthworm City, Illinois.

DEAR HENDERSON: It gives me great pleasure to report that I have started my activities in this town with even more than my usual energy. Although I arrived only this morning, I have already become deeply involved in the promotion of a gold mine.

In case you should wonder why a man who was sent out primarily to infuse new life and selling enthusiasm in the Earthworm-tractor dealers of the country should turn to mining, I will explain that this new activity is necessary in order to stimulate and revivify Mr. George Dow, who is our dealer in this territory, and also one of the laziest yokels I have ever run across.

Honestly, Henderson, he is as sleepy, as dumb and as contented as a superannuated Holstein cow. When I called on him this morning at his office, he languidly invited me to take a chair and then sat down himself, arranged his long, skinny legs on the desk, stuck an old pipe in his mouth and settled himself as if he was going to take a nap. He was too lazy even to light the pipe.

"How's business?" I asked.

"Can't complain," he said.

"Selling any tractors?"

"No."

"Got any on hand?"

"Yes. I got two sixty-horsepower models in the warehouse."

"How long since you've sold a tractor?"

"I don't know. I guess the last sale I made was about a year ago—maybe a year and a half."

"That's terrible," I said. "I had no idea things were so bad up in this country. But you can cheer up, Mr. Dow, for the Earthworm Tractor

**He languidly invited me to take a chair and then sat down himself,
arranged his long, skinny legs on the desk.**

Company has sent me out to help you dealers. I am ready to assist you in
any way I can, and if we get together and attack this situation with energy
and enthusiasm, I feel sure we can achieve some real results."

At this, Mr. Dow smiled in what might be described as an amused and
tolerant manner. He was a rather likable old bozo, maybe sixty years old,
and he had a lot of pleasant-looking little wrinkles around his eyes.

"Well, well," he said, "it does beat all how you young chaps are always
rushing around, full of energy and excitement. But when you're as old as I
am, my boy, you will learn to take things more calmly."

"Possibly," I said, "you are taking things too calmly. If you don't get
out and buzz about a bit, how can you expect to pull yourself out of this
depression?"

"But there isn't any depression here," he said. "In Vermont we don't have
these crazy booms that you have in the big cities, and we don't have big
depressions either. We just live along quiet and sensible. At the moment
it is true that business is rotten and times are rather hard. But it doesn't
bother me at all."

"What?" I asked. "Business is rotten, and it doesn't bother you?"

"Certainly not. I've got everything I need. I have a nice house at the edge
of town here, a splendid vegetable garden, a couple of cows, a fine flock of
hens, a little orchard and a woodpile big enough to last all next winter. I
have about five thousand dollars in the bank—and I intend to keep it there.
Most of my neighbors are fixed the same way, so none of us have to worry

very much about hard times. Of course, there are a few people who are out of luck and have nothing at all, but we have them with us even when times are supposed to be good. In some ways I am better off than the average."

"I thought you just said you hadn't sold a single tractor for over a year."

"Yes, but I have a little garage and filling station that I run in connection with my tractor agency, and this business is still going on—after a fashion."

"Is it making any money?"

"No, but it's not losing any, either. And I'm paying the help enough to live on. Another fortunate thing is that my son, who has been taking a course in mining engineering down at Columbia, has just graduated, so I won't have any more expense for his college course. He hasn't been able to get a job, so I have told him to come home and stay with his father and mother, where the cost of living is almost nothing. He will be here in a day or two. There are plenty of good chairs in the office here, and room enough for another pair of feet on the desk, so he and I can just sit around and discuss the weather and politics and so on, and thumb our noses at this much-advertised business depression."

Having finished these remarks, Mr. Dow stretched himself contentedly and settled down even deeper into his comfortable swivel chair. It seemed to be up to me to make some adequate reply. And this was difficult, because, although Mr. Dow's line of argument was completely cockeyed, it nevertheless had a certain plausibility about it. If he actually had everything he needed already, and was completely satisfied, possibly there was no logical reason why he should start tearing around and trying to rake in more cash. For a moment he almost had me converted to his point of view. But, naturally, a high-powered salesman like myself, whose whole business is to inspire pep and energy, cannot afford to listen to any such insidious propaganda as Mr. Dow was putting out.

"Your whole point of view is wrong," I said. "To hear you talk, anybody would think you had given up even trying to sell tractors."

"Of course I have," he said. "Why should I waste my time trying to sell tractors when there isn't anybody around here at the present time that needs any such thing?"

"But there must be a few possible customers. How about the farmers?"

"They are all feeling too poor."

"Any lumbering going on in these mountains?"

"There used to be, but there isn't any more."

"Any industries that could use tractors?"

"No. We used to have a pulp mill and several sawmills. And a long time ago we even had a gold mine. But they all shut down."

"You say there used to be a gold mine? I didn't know they had such things in Vermont."

"Oh, yes," said Mr. Dow. "There has always been a certain amount of mining in the Green Mountains—iron, copper, gold, silver and so on. But they're not doing anything these days—can't compete with the big Western production. This particular gold mine belongs to me."

"It does?"

"Yes. I inherited it from my father. He opened it up back in the '70s, and sank about all the money he had in it. If you look out the window there, you'll see a little tumble-down shed up on the side of the mountain. That shed is at the mouth of the mine."

"I see it," I said. "Is there really any gold there?"

"Oh, yes. There's a little. But not enough so my father could make it pay. He went completely busted."

"And you've never tried to work it yourself?"

"Absolutely not. I don't know anything about mining, and I don't want to. As I told you before, I'm perfectly happy and contented, and I intend to remain so. I'm not going to risk any of my money in any speculative enterprise, whether it's gold mining or high-powered tractor-selling campaigns, or anything else."

"Well," I said, "maybe you're right. But don't get the idea that a tractor-selling campaign necessarily takes a lot of money. The important thing is plenty of energy and persistence. There must be some place around here where we can sell tractors. Is there any road building going on?"

"Yes, the local road commissioner is putting in several miles of driveway through a tract of woods that has been bequeathed to the town as a park. It's quite a big job. There's a lot of grading and cut-and-fill work. The commissioner is spending about fifteen thousand dollars—which was left for this purpose by the man who bequeathed the park."

"Is the commissioner using tractors?"

"No."

"All right. We'll sell him a few."

"We can't. He's doing the whole thing by hand labor—picks and shovels and wheelbarrows."

"I'll soon argue him out of that idea," I said. "Just wait till I show him how much labor he'll save by getting rid of his medieval implements and doing the job as the Lord intended—with Earthworm tractors."

"But he doesn't want to save labor."

"Why not? Is he crazy?"

"No. He's just trying to be public-spirited. He is building the road without machinery so as to provide work for all of the town's unemployed who want it—which means about twenty men. Besides helping these people, the commissioner feels he is benefiting the whole town by keeping all the money right here."

"A swell benefit that is to the town as a whole," I said. "All that guy is doing is subsidizing inefficiency. He may be handing out a little temporary relief to a few unfortunates, but he is cheating all the rest of the town by giving them only about a quarter as much road as they are entitled to from the money he is spending."

"Maybe you're right," said Mr. Dow, "but you'll never make the road commissioner see it that way. He has made up his mind that he is the guardian angel of the unemployed, and nobody can tell him any different. He's just as stubborn as I am."

"In that case," I said, "he must be a tough customer. But there must be some way to get to him."

"I'm afraid not," said Mr. Dow. "He'll never buy a tractor as long as there are men in this town who want work and can't get it."

At this point I suddenly got one of those keen and practical ideas which so often come to me in difficult situations. "I know how we can handle this thing," I said. "It is very simple. All we have to do is get rid of the unemployed."

"Drown them in the river or something?"

"No. Give them jobs."

"Doing what?"

"Anything. You might, for instance, open up your gold mine and hire all these men away from the road commissioner. Then he would have to buy a tractor."

"Maybe so. But where would I come out? I'd be paying wages to twenty men and nothing to show for it."

"You would have a lot to show for it," I said. "You would sell a four-thousand-dollar tractor, on which your commission would be eight hundred dollars. That would carry your pay roll for a week and still show you a handsome profit. At the end of the week—after you had sold the tractor—you could shut down."

"And turn all those men loose again without any jobs?"

"Oh, they would be all right," I said. "The road commissioner would probably hire them back. He has enough money to buy a tractor and hire these men besides. But you might find you could make enough out of the mine to keep running for quite a while."

"And eventually go bust like my father did."

"No," I said. "If you were a deep student of political economy, such as I am, you would know that this is a peculiarly auspicious time to run a gold mine."

"In spite of the depression?"

"Because of the depression," I said. "Production costs—labor and supplies and everything—are low. And the product, gold, automatically sells at par. You would have a much better chance than your father had."

"Well, I don't know," said Mr. Dow. "I'm not a student of political economy, like you, but I do know that in the '70s, when my father was working this mine, prices were even lower than they are today. And gold was selling at a premium, because all the money was paper, which was not redeemable in specie. It looks like my father had a better chance than I would have."

"At first sight," I admitted, "it might look that way. And as you are not a student of political economy, I can't go into the matter deeply enough to show you where you are wrong. But, regardless of theoretical considerations, I am convinced that it would be of tremendous practical benefit both to yourself and to the tractor business if you would take a chance on this gold-mining proposition."

"I wouldn't even consider it," said Mr. Dow. "I've told you before that I don't know anything about gold mining—and I don't want to know anything about it."

"You said your son has just graduated in mining engineering, and has no job. Why not let him run this mine?"

"And bankrupt the whole family? I should say not."

"Your son will be home in a day or so, won't he?"

"Yes."

"It wouldn't hurt anything to talk the thing over with him?"

"No, I suppose not."

"All right then," I said, "I won't bother you anymore about it right now, but when your son arrives we will get together and decide what to do."

"It don't sound good to me," said Mr. Dow.

"Perhaps you will change your mind later," I said. "And now I think I'll be getting back to the hotel for lunch. I'll see you again this afternoon or tomorrow. Goodbye."

I left the old guy sprawled out languidly in his office chair with his feet in the same position on top of the desk. As I walked along toward the hotel I kept turning this gold-mine idea over and over in my mind, and the more I thought about it the better it seemed. By the time I reached the hotel, I had made up my mind to make a little private investigation of the mine as soon as possible. As I am fairly ignorant about such matters. I decided I would need expert advice. On the chance that I might possibly run across something, I asked the hotel clerk whether there was such a thing as a mining engineer in town, and if not, where would be the nearest place I could find one.

The clerk did not even seem to know what a mining engineer was. He doubted if any such thing could be found in town, or even in Rutland or Burlington. He said, however, he would make inquiries. After thanking him I went in to lunch.

Half an hour later, when I came out of the dining room, the clerk called me and introduced a very courteous and pleasant-spoken gentleman by the name of Mr. Bailey Bryant. It appeared that Mr. Bryant was from New York, and had arrived at the hotel two or three days before to take advantage of the excellent fishing in the mountain streams of the neighborhood. He had heard the hotel clerk inquiring for a mining engineer, and as that was his profession he had stepped forward to offer his services.

This was indeed a piece of splendid good luck, and a striking example of how fortune seems to favor a man like myself who is always on the alert and ready to grasp any opportunity that presents itself.

At once I took Mr. Bryant aside, explained the whole situation completely and frankly, much as I have set it forth in this letter, and asked him if he would look over the mine and give me his opinion on it. He replied that he would be delighted, and we started to walk up the mountain to the little shed which Mr. Dow had pointed out to me.

On the way Mr. Bryant gave me a number of fascinating reminiscences of his gold-mining experiences in South Africa, in California and in Alaska. Seldom have I met a more brilliant conversationalist. As I am a very good judge of men, it did not take me long to realize that in spite of his modest manner Mr. Bryant possessed an unusually powerful intellect and was a real authority in his field.

When we reached the mine we discovered that it was a small tunnel, perhaps six feet in diameter, cut horizontally into the solid rock of the mountain. Mr. Bryant, with admirable foresight, had brought an electric flash light. Although the tunnel was partially obstructed by fallen rocks

Mr. Bryant gave me a number of fascinating reminiscences of his gold-mining experiences in South Africa, in California and in Alaska.

and rubbish, we had no trouble in penetrating to its end, which was per-haps a hundred yards from the opening.

My companion subjected the rock walls to the most minute and pains-taking examination, while I stood by in a state of suppressed excitement—eager to ask questions, but holding on to myself so as not to interrupt the train of his thought. From time to time he loosened small fragments of stone, examined them with a lens and put them in his pocket. He seemed particularly interested in the rock at the end of the tunnel, and here he took three or four samples. After perhaps half an hour, we returned to the opening.

"This is most interesting," he said. "As a careful scientist, I naturally hesitate to commit myself absolutely, but I do not mind telling you that every indication points to the fact that the work on this mine was stopped at the very moment when the richest ore was about to be uncovered."

"You really think so?" I asked.

"That is my opinion," he said. "Look. Here is a specimen from about the middle of the tunnel. Ordinary Barre granite with a trace of mica. Possibly a little gold in it, but obviously not much."

"It certainly is not much to look at," I said.

"No," he agreed, "but now cast your eye on this, which came from the end of the tunnel." He held up a piece of whitish stone. "Pure quartz, and in a formation completely analogous to that of the great Comstock Lode in Nevada. It is most interesting."

"Then you really think it would pay to work this mine?"

"Before giving you a definite answer, I should have to make a more complete survey. It will be necessary for me to blast off a little more rock at the end of the tunnel, so as to uncover the richer ore which I hope to find just back of the present face. Then, after making a chemical and microscopic examination of the ore, I would be in a position to give you a very definite opinion based upon solid scientific facts."

"Would you be willing," I asked, "to make this examination for me?"

"Certainly," he replied. "You understand, of course, that I will have to charge you for it. I have been glad to give you my unofficial opinion free, but for a regular professional report I would have to ask my regular professional fee, which is a hundred dollars."

"Holy Moses!" I said. "A hundred dollars?"

"That," he said, "is my regular fee."

"Well," I said, "I guess that's perfectly reasonable for expert advice from an important guy like you, but I don't know just how I am going to pay it. I can't afford the expense myself. And I'm afraid the company wouldn't let me get away with it on the expense account. Mr. Dow really ought to pay it himself. But he's so tight with his money that I don't believe we could talk it out of him—unless, of course, it turned out that the mine was a real bonanza like you think it is. And listen, that gives me an idea. How would you like to gamble on this—double or quits?"

"What do you mean?" he asked.

"If it turns out that this thing is a big money maker. I'll guarantee you two hundred dollars—double your fee. In that case I'm practically sure I can get that much out of old man Dow. Almost anybody would pay two hundred dollars for a real gold mine. So I'm willing to take a chance, and, if he won't pay you, I will. On the other hand, if we don't find this rich ore you're looking for, we'll call it quits and you get nothing. How about it?"

"Fair enough," said Mr. Bryant. "Every mining engineer is a gambler at heart, and I'm no exception."

We shook hands on the bargain and returned to the town, where Mr. Bryant left me. He said he would get the supplies he needed and return to the mine to make his investigations, and he courteously declined my offer of assistance, saying that he could work better and concentrate more effectively when he was alone.

I spent the remainder of the afternoon seeing if I could discover any further opportunities for possible tractor sales. I ran into nothing, however, that looked in any way promising.

This evening I have been writing this report. I have just seen Mr. Bryant in the lobby. He reports that his investigations have proceeded very satisfactorily, and he wishes me to bring Mr. Dow up to the mine tomorrow morning so that we can all look over the ground together. When I asked him whether his report would be as favorable as he had hoped, he merely smiled and said that he would prefer to say nothing until tomorrow.

I have just called up Mr. Dow and arranged to go with him to the mine in the morning. So the stage is all set, and I have high hopes.

I have given you this very full account of my activities, so that you can see I am pulling off a very remarkable and brainy piece of work here, and so that there will be no question about allowing the two hundred dollars in case I should have to put it in my expense account.

Very sincerely,
ALEXANDER BOTTS.

———

ALEXANDER BOTTS
SALES PROMOTION REPRESENTATIVE
EARTHWORM TRACTOR COMPANY

Dow's Gulf, Vermont.
Tuesday, June 7, 1932.

Mr. Gilbert Henderson,
Sales Manager,
Earthworm Tractor Company,
Earthworm City, Illinois.

DEAR HENDERSON: What a day this has been—what a day! When I wrote you yesterday, I was expecting something pretty good, but never in my wildest dreams had I anticipated anything so sensational and so glorious as what has actually come to pass. But I must try to begin at the beginning and relate my wonderful news in an orderly manner.

This morning Mr. Bryant, Mr. Dow and I walked up to the mine. Mr. Bryant maintained his quiet and sophisticated reserve, but there was about him a subtle air of optimism which both stimulated me and aroused my curiosity. Mr. Dow was good-natured, but inclined to scoff at the whole expedition. I had told him that my friend was a mining engineer who had kindly consented to look over his mine and give him an opinion as to its value. As for myself, I was in such a flutter of anxiety and suspense that I could hardly wait to hear Mr. Bryant's verdict.

When we reached the mouth of the tunnel, Mr. Bryant gave us a brief description of the geological formation in which the mine was located, and then explained the various factors which had caused him to believe that there might be a very rich vein of what he called auriferous quartz immediately beyond the end of the tunnel. This explanation was couched in scientific language—which I will not repeat, partly because I can't remember it, and partly because you guys at the tractor factory couldn't understand it anyway. After giving us the lowdown on these theoretical matters, Mr. Bryant told us what he had done the previous afternoon.

"I got some tools and a couple of sticks of dynamite at the hardware store," he said. "Then I came up here, drilled a hole into the rock face at the end of the tunnel, and set off a small charge. The results"—and here he smiled in a mysterious and highly dramatic manner—"were so favorable that I think they can be appreciated by almost anyone. If you will step this way I will show you something that you will have no trouble in comprehending, no matter whether you have any technical training or not."

Snapping on his electric flashlight, he led us through the tunnel to its extreme end. At once I noticed that the blast he had spoken of had exposed a fresh surface of rock. It was white quartz.

"Look at it a little closer," said Mr. Bryant.

Mr. Dow and I did so. And what I saw made my heart start beating and pounding in a way that it has not done for years. All over the surface of the rock, embedded in the seams and crevices, were thousands of glittering, shining yellow particles. Some were as small as grains of the finest powder. Others were the size of peas, and a few were actually as large as small acorns. It was some minutes before I could speak.

"Is it," I asked, "is it really gold?"

"Yes," said Mr. Bryant, "it is gold—and as remarkable a vein as I've seen for years." He turned to Mr. Dow. "Permit me to congratulate you," he said. "You are many times a millionaire."

It was quite evident from Mr. Dow's expression that this great discovery had knocked his intellect into a confused and dizzy whirl. He leaped

"Look at it a little closer," said Mr. Bryant. What I saw made my heart start beating and pounding in a way that it has not done for years.

forward and began clawing little chunks of the beautiful yellow metal out of the rock and turning them over and over in his hands. His naturally cautious nature prevented him from completely accepting the obvious facts, but at the same time, he was rapidly becoming drunk at the sight and the feel of the gold, and at the thought of the incredible riches that now were his.

"I can hardly believe it," he kept saying. "I can hardly believe it. Is it really true that I am a millionaire?"

"Well," said Mr. Bryant, "the great Comstock Lode yielded something over three hundred million dollars."

"Three hundred million dollars!" gasped Mr. Dow.

"Three hundred million dollars," repeated Mr. Bryant. "And the geologic formation here is strikingly similar to the Comstock formation—which is a fact of great scientific interest. Of course, I do not claim that this vein will turn out as well. It might yield much more, but, on the other hand, it might not run more than fifty or a hundred million."

"Well," said Mr. Dow, "even if it was only one million dollars, that would be quite a bit of money." He continued clawing out little pieces of the metal. Apparently, he couldn't wait even a moment, and he wanted to get the stuff into his hands as fast as possible.

I began to get the fever myself, and it occurred to me that possibly I had handled this thing wrong. The investigation of the mine had been my idea in the first place, and with a little more forethought I might have managed so that I would have owned a substantial share.

Mr. Bryant, the engineer, was the only one of us who maintained a normal calmness. "I would suggest," he said to Mr. Dow, "that you say nothing about this to anyone until you have checked over your title to the land and made absolutely sure that no one else has even a shadow of a claim on it. Furthermore, you may want to buy additional land, in case investigation should indicate that you do not own the entire deposit."

"That certainly is a good idea," said Mr. Dow. "You have a real head on you, Mr. Bryant. How would you advise me to go about working this mine?"

"That," said Mr. Bryant, "will have to be decided after a thorough investigation of all the factors involved. Unfortunately, I won't be able to stay long enough to see the mine put on a permanent operating basis. I'm up here only for a short vacation, and pressing business will make it necessary for me to return to New York very soon. But, if you wish, I can stay around for three or four days to advise you and get you started right."

"That would be fine," said Mr. Dow. "That would be wonderful. You don't know how I appreciate this, Mr. Bryant. How can I ever repay you for what you have done?"

"Well," said Mr. Bryant, "my fee for this preliminary examination, as agreed upon by Mr. Bolts and myself, is two hundred dollars. That will include my time for yesterday and today. For each additional day, starting tomorrow, I will charge you my usual rate of one hundred dollars a day."

The mention of this rate of payment served to arouse momentarily Mr. Dow's deeply ingrained New England thrift. "A hundred dollars a day is a lot of money," he said.

"Yes," agreed Mr. Bryant pleasantly, "but not a great deal in comparison to a three-hundred-million-dollar gold mine."

"You're right," said Mr. Dow, "and you will have to excuse me for seeming so small-minded. As a matter of fact, I guess your charges are pretty small, and I will be delighted to pay them. They will be a mere drop in the bucket as compared to the gold in this mine. But I was just wondering—have you really analyzed this stuff in here? Are you absolutely sure it's actually gold, and not just copper or something? Wouldn't it, perhaps, be a good idea to get some real chemist to analyze it for us, so we'd be certain? I have a cousin that teaches in the chemistry department of Middlebury College, which is only about thirty miles away across these mountains. I could take a sample over and he could tell me all about it."

"If you ask me," I said, "I would say that you would just be wasting your time. Mr. Bryant is a professional mining engineer. He has worked in mines all over the world. And he knows his gold, and his silver, and his copper, and his tin and everything else. If he says it's gold, that's what it is. What more could you learn from a fool college professor?"

At this point Mr. Bryant spoke up and showed what a truly broad-minded man he is. "Mr. Dow is perfectly right," he said. "To a man of my experience, it is evident that this material is gold. It is my business to know about such things, and I do know. I am certain. But it is perfectly natural for a man like Mr. Dow to have his doubts. And it is reasonable, in a matter of so much importance, that he should check the facts carefully. I should very much prefer that he get an independent opinion. In fact, I was going to suggest it myself. Here, take this."

He reached into his pocket, drew out a small pasteboard box and handed it to Mr. Dow.

"What's this for?" asked Mr. Dow.

"It is something to carry the gold which you have been pulling out of that rock."

By this time Mr. Dow had a whole handful of small nuggets. He put them in the pasteboard box. Mr. Bryant snapped a rubber band around it and handed it back.

"There you are," he said. "My advice would be that you step into your car, drive over to Middlebury and get your cousin to analyze that gold right away. While you are gone I will start an examination of this entire mountainside with a view to determining the possible extent of this deposit of auriferous quartz."

"It's a good idea," said Mr. Dow. "I'll start right away."

"And remember," cautioned Mr. Bryant, "don't breathe a word of this to anyone. Don't tell your cousin any more than you have to, and impress it upon him that the matter is to be kept secret."

"You can trust me," said Mr. Dow.

We all returned to the town, and Mr. Dow drove away in his ear. Mr. Bryant invited me to accompany him on his geologic expedition, explaining laughingly that he was afraid to leave me sitting around the hotel lobby, for fear I might be tempted to spread the great news. We had a very interesting day, tramping about the mountains. As I have said before, Mr. Bryant is one of the most charming conversationalists I have ever met. And the information which he gave me about the various rock formations was not only educational but highly interesting and enjoyable.

We got back to the hotel late in the afternoon, and soon afterward Mr. Dow arrived from Middlebury. He seemed even more agitated than in the morning, and as soon as the three of us were alone in my room he actually began dancing around the floor, gibbering at us like an idiot.

"It's all true!" he said. "My cousin analyzed it, and it's gold! Think of it! Real gold! And there's millions of dollars' worth in that mountain! And it all belongs to me! It all belongs to me!"

The old guy raved along, gloating over his good fortune in such a disgusting manner that I could not help being slightly annoyed—especially as I was to have no share in all this money that he was yawping about so loudly. At the same time I could get a little cynical amusement out of the change that had taken place in Mr. Dow's character. He had completely lost that quiet, philosophical manner, that quiet contentment with the simple life, which had been so evident at our first meeting. The sudden appearance of this vast amount of filthy gold had turned him into as greedy and material-minded a man as I had ever met. Ah, well, I thought to myself, all these things are no direct concern of mine. I came here for the purpose of promoting a few tractor sales, and it looked as if I were going to do it.

After a short conference we decided upon a definite plan of action for the next few days. Mr. Bryant is going to write to various firms in different parts of the country, asking their prices on machinery to be used in working the mine. The ore is so rich that the mine could be run at a profit with the crudest possible equipment, but Mr. Bryant pointed out that the greatest profits could only be realized through the use of the most up-to-date machinery. While waiting to hear from the manufacturers, Mr. Bryant will continue his geologic investigations.

Mr. Dow is going over tomorrow morning to hire away practically all the road commissioner's workmen. With these men and one of the two tractors which he has on hand, he is going to start building a road to the mine so that the heavy ore-crushing machinery, when it arrives, may be moved up without delay. He will tell his men that the road is to be used for getting out logs, and will carefully guard the secret of the great gold strike. As for myself, I will take the second of the two tractors which Mr. Dow has on hand, and start giving a demonstration to the road commissioner.

Thus you see that everything promises to work out splendidly. I am still somewhat confused by the magnitude of the events which have been transpiring, but I trust I have given you a reasonably clear account of them.

Kindly keep all this news about the gold mine strictly confidential. We do not wish the discovery to leak out at this time.

Very sincerely,
ALEXANDER BOTTS.

———

ALEXANDER BOTTS
SALES PROMOTION REPRESENTATIVE
EARTHWORM TRACTOR COMPANY

DOW'S GULF, VERMONT.
THURSDAY NIGHT, JUNE 9, 1932.

MR. GILBERT HENDERSON,
SALES MANAGER,
EARTHWORM TRACTOR COMPANY,
EARTHWORM CITY, ILLINOIS.

DEAR HENDERSON: Two days have passed since my last letter and there is nothing much to report except that everything is going along as planned. The great gold discovery is still a secret. Mr. Bryant is continuing his geologic researches. Mr. Dow is hard at work on his new road. He has two lawyers searching the titles of his land. And he is so smug and self-satisfied that he is positively obnoxious.

As for myself, I have been spending the last two days demonstrating the tractor to the road commissioner. This gentleman is a little sore at the high-handed way in which Mr. Dow has hired most of his men away from him. But he is a reasonable guy, holds no grudge against me, and is so favorably impressed with the tractor that I think he will decide to buy it in a day or so.

No more at present.

Very truly,
ALEXANDER BOTTS.

ALEXANDER BOTTS
SALES PROMOTION REPRESENTATIVE
EARTHWORM TRACTOR COMPANY

DOW'S GULF, VERMONT.
FRIDAY, JUNE 10, 1932.

MR. GILBERT HENDERSON,
SALES MANAGER,
EARTHWORM TRACTOR COMPANY,
EARTHWORM CITY, ILLINOIS.

DEAR HENDERSON: I regret to report that affairs here are not progressing in exactly the way I had expected. In fact, it now appears that in certain important matters I was so completely and absolutely mistaken that it is most embarrassing for me to explain what has occurred. However, it is always my policy to be perfectly frank. So in this case I will go ahead, following my usual custom, and tell you the whole distressing truth.

I spent this morning and most of this afternoon uneventfully enough, demonstrating the tractor on the road commissioner's cut-and-fill job. The commissioner himself was not present, being in town on some other business, but I went ahead just the same. At five o'clock I returned to town and stopped in at the post office to buy some stamps.

And here, without any warning whatsoever, I received what I can only describe as a very severe shock. As I turned away from the stamp window my eye was suddenly attracted to a picture which was posted on the bulletin board. This picture was a printed reproduction of a photograph, and it showed with great clearness the handsome and distinguished features of Mr. Bailey Bryant, the mining engineer from New York.

Above the picture were the words, "Wanted, For Grand Larceny, Uttering Counterfeit Notes, Using the Mails to Defraud, Obtaining Money Under False Pretenses, and Forgery." Beneath was a long list of names: "John Bailey, alias John Bryant, alias Bailey Spencer, alias Spencer the Spieler, alias Peter Livingston, alias Pete the Prestidigitator." After a full physical description, there followed further information: "This man is well-educated, is a good talker on almost all subjects, has a most pleasing personality, and is remarkable for his versatility and for the variety of his swindles. Is an accomplished amateur magician and has, on several occasions, robbed jewelers by substituting an empty box for one containing valuable gems. Has posed at various times as physician, clergyman, Army officer and college president." The notice ended with a request that anyone apprehending this person should place him under arrest and communicate with a gentleman called Mulrooney down in New York.

You can well imagine the astonishment and consternation which filled my mind. At once I removed the notice from the bulletin board, folded it up, put it in my pocket and hurried over to Mr. Dow's office. Mr. Dow greeted me and introduced me to his son, who had arrived the day before from Columbia.

"Mr. Dow," I said, "I have very grave news for you. I have just discovered certain facts which lead me to believe that our friend, Mr. Bailey Bryant, is not the sort of man we think he is. In fact, I'm beginning to suspect that there may be something faintly cockeyed about this whole gold-mining enterprise."

"Yes," said Mr. Dow, "the same thought occurred to my son and myself this morning."

"This morning?" I asked.

"Well, my son began to be suspicious as long ago as yesterday afternoon. As soon as he got home I told him the great news, but he was not so enthusiastic as I had expected. It seems that last fall, without telling me anything about it, he had examined the mine, gathered up a few samples of the ore, and taken them to New York with him to be analyzed. The analysis showed a certain amount of gold, but there was nothing to indicate even the possibility of any such rich deposits as we had run across. So yesterday my son went up to the mine, collected a few of those little golden nuggets and made some tests on them. They turned out to be nothing but brass."

"This is terrible," I said. "Did you tell Mr. Bryant about it?"

"At that time," said Mr. Dow, "Mr. Bryant was up in the mountains somewhere on one of his geological expeditions. In his absence we made a few inquiries around town."

"And what did you find?"

"We found that he had bought some brass pipe and a cheap shotgun at one of the hardware stores. Then he had visited a blacksmith and had him melt up the brass pipe and pour it out so that it formed a lot of small, irregular globules. He had also paid a visit to a dentist."

"You mean he was having trouble with his teeth?"

"No, he bought five dollars' worth of gold, and had the dentist melt it up and pour it out in little chunks very similar to the brass which the blacksmith had worked on. All this we found out yesterday. This morning we got Mr. Bryant in the office here and told him that we had the goods on him, and his game was up."

"What did he say?" I asked.

"At first he was very smooth, and pretended he didn't know what we were talking about. But when he saw that we had him right and proper, he owned up."

"He explained everything to you?"

"Yes," said Mr. Dow. "And some of it was pretty good. Especially what he had to say about you."

"Good Lord," I said, "what did he say about me?"

"He said that up to the time he met you he hadn't intended to pull anything crooked in this town at all. He was here partly for some fishing and a little vacation, and partly to get away from what he described as a rather unfortunate situation down in New York. But when he saw you he said you were such a perfect example of a born easy mark that he couldn't resist taking you for a ride."

"Well," I said sadly, "he certainly succeeded. He took me on a whole cruise. And I'm not sure that I understand even yet just how he managed everything."

"He told us," said Mr. Dow, "that he was going to be paid only in case the mine turned out to be worth something. That is why he arranged such a spectacular set up. He wanted to make the mine look as good as possible. After he had uncovered a new face of rock with dynamite, he shot all that brass into it with his shotgun."

"But what about that sample that your cousin analyzed?"

"Apparently, Mr. Bryant's hand is quicker than my eye. While he was putting a rubber band around the box I had filled with brass, he slipped it up his sleeve or something, and then handed me back another box with the dentist's gold in it."

"And to think," I said, "that I actually believed that guy, and swallowed everything he told me. I certainly owe you my most abject apologies. I thought I was helping you, and all I have done is to put you to a lot of trouble and expense over a mine that isn't worth a nickel."

"It isn't as bad as that," said Mr. Dow's son. "The mine is no such bonanza as you thought. But the analysis of the ore, which I made in New York last winter, shows that there actually is a small amount of gold present. In my grandfather's day the mine was a failure. But now, by putting in a little modern machinery and using the cyanide process, I'm certain that we can work it with a small but steady profit. I had intended to ask my father to finance the thing and let me run it, but he is so cautious that he never would have consented unless you and Mr. Bryant had got him started. As it is, he has to go on. He can't back out now."

"Why not?" I asked.

"He has already built a road halfway up to the mine. And he promised all those men he hired from the road commissioner that he would give them work for the rest of the year. So he has to go on—and it's a very good thing."

I looked at Mr. Dow, Senior, to see what he thought of this. He was all stretched out in his office chair with his feet up on the desk, and his whole attitude expressed the same languid contentment which had been so noticeable on the day I first met him.

"Yes," he said, "the whole business suits me fine. For a while you had me all excited and het up over the idea of being a millionaire. And that is all wrong for a man of my age who has always been used to peace and quiet. If the excitement had kept up much longer I would have just naturally worn myself out and come to an unfortunate and premature end. As it is, I can sit around as usual with nothing in the world to worry me. On the other hand. I'm grateful to you because—"

"Why should you be grateful to me?" I asked.

"Because I had perhaps been getting a little too quiet and contented, and you have shaken me up enough to get me started on this gold mine. It will make a very nice little business for my son. It has already provided employment for a number of people here in town. And besides that, it has been of some assistance to the tractor business. I saw the road commissioner this afternoon and he told me he wants to buy the machine which you have been demonstrating. So everything is fine, and my thanks are due to both yourself and Mr. Bryant."

"By the way," I said, "what did you do with that guy? You didn't pay him anything, did you?"

"No. He agreed with us that, under the circumstances, we owed him nothing."

"But I suppose you know he's wanted down in New York." I showed him the notice I had brought from the post office.

"Well, well," said Mr. Dow. "My son and I had not seen this. If we had known about it, I suppose it would have been our duty to turn him over to the police. But as it was, we had no grudge against him and we let him go. He took the noon train for Canada. And on the whole, I'm glad he got away. It was a lucky thing for us he came to town. And I can say the same thing regarding you, Mr. Botts."

"I hope," I said, "that you're not classing me with Mr. Bryant. Certainly, you don't believe that I'm a crook?"

"Oh, no, indeed," said Mr. Dow. "Anybody as slow-witted as you couldn't be a crook. We realize perfectly that you are completely honest and that you meant well. And as a matter of fact you have, in your awkward way, helped us and our business a great deal. So we thank you from the bottom of our hearts."

As I could not think of any good answer to these various cracks of Mr. Dow's, I wished him and his son good afternoon as politely as I could and returned to the hotel. I am leaving tomorrow for the next job.

As ever,
ALEXANDER BOTTS.

THE TRACTOR BUSINESS IS NO PLACE FOR WEAKLINGS

ILLUSTRATED BY TONY SARG

EARTHWORM TRACTOR COMPANY
EARTHWORM CITY, ILLINOIS
OFFICE OF THE SALES MANAGER

TUESDAY, FEBRUARY 2, 1932.

MR. ALEXANDER BOTTS,
TYLERVILLE HOTEL,
TYLERVILLE, WISCONSIN.

DEAR BOTTS: According to your schedule, you should be arriving in Tylerville day after tomorrow. In a former letter, you will remember, we requested you to call on our dealer in Tylerville, Mr. A. H. Smith, of the Smith Tractor Company, Incorporated, and assist him in putting on an intensive selling drive. Recent developments, however, make it necessary for us to alter these instructions.

Mr. Smith has written us that the building which housed his business has burned, and that his affairs are in very bad shape. His contract as dealer for the Earthworm Tractor Company expires in two weeks, and he says he will be unable to renew it.

An implement dealer by the name of John Yerkes, in Tylerville, has written us that he would be glad to take over the dealership which Mr. Smith is giving up.

Instead of consulting with Mr. Smith, therefore, you will make a thorough investigation of Mr. Yerkes and any other possible candidates for the dealership, and sign a contract for the coming year with the man who seems to you the most promising.

Incidentally, I may say that Mr. Yerkes's letter has impressed me most favorably. He sounds like a good practical man with plenty of force and energy.

I enclose several copies of Form CX-447-G—our standard dealer's contract.

Very sincerely,
GILBERT HENDERSON,
Sales Manager.

ALEXANDER BOTTS
SALES PROMOTION REPRESENTATIVE
EARTHWORM TRACTOR COMPANY

TYLERVILLE HOTEL,
TYLERVILLE, WISCONSIN.
THURSDAY EVENING, FEBRUARY 4, 1932.

MR. GILBERT HENDERSON,
SALES MANAGER,
EARTHWORM TRACTOR COMPANY,
EARTHWORM CITY, ILLINOIS.

DEAR HENDERSON: I arrived in town this afternoon and found your letter awaiting me. Right away I got busy. I inquired at the local chamber of commerce and elsewhere, and found that Mr. Yerkes is the only man in town who is in a position to take over and successfully carry on the dealership which Mr. Smith is giving up.

Before offering Mr. Yerkes the job, however, I decided I would have a look at this Smith guy. I had met him once several years ago, and I remembered him as one of our most successful dealers.

"Perhaps," I said to myself, "old man Henderson's idea that we ought to drop him in the garbage can this way is just as cockeyed as some of his former ideas." So I went around and called on Mr. Smith. I asked him what was the matter. And I am very glad that I did.

The poor fish has certainly been having rotten luck. His business is completely bankrupt, and he himself seems on the verge of a nervous breakdown. Honestly, Henderson, his story was pitiful. The way he talked about his dear old grandfather almost had me in tears. And before I knew what I was doing, I had promised to help him. At first sight, the case appears hopeless. To meet his immediate obligations he will have to borrow fifty thousand dollars on no security. And this isn't all. If he borrows the money, he will have to pay it back sometime, and it doesn't look to me as if he ever could.

But difficulties such as these never discourage Alexander Botts. I am leaving this afternoon for Chicago, where I will attempt to negotiate the necessary loan. If I fail, which is more than likely, the Earthworm Tractor Company will have to advance the fifty thousand dollars. Otherwise we can't keep Mr. Smith as our dealer.

I am sure that you will be delighted to know about this generous piece of rescue work. My address in Chicago will be the Blackstone Hotel.

Most sincerely,
ALEXANDER BOTTS.

———

EARTHWORM TRACTOR COMPANY
EARTHWORM CITY, ILLINOIS
OFFICE OF THE SALES MANAGER

FRIDAY, FEBRUARY 5, 1932.

MR. ALEXANDER BOTTS,
BLACKSTONE HOTEL,
CHICAGO, ILLINOIS.

DEAR BOTTS: Your letter is received, and I am not in any way delighted to know about "this generous piece of rescue work." I don't wish to seem harsh or unfeeling. And I deeply regret the fact that Mr. Smith's business has failed. But you must realize that the Earthworm Tractor Company cannot pay your salary and expenses while you rush about the country attempting to negotiate a large unsecured loan for a completely bankrupt gentleman who has already notified us that he has severed his business connection with us. Your suggestion that the Earthworm Company might lend Mr. Smith fifty thousand dollars is absurd. This company is a business concern, not a philanthropic organization.

You will return to Tylerville at once, and follow out my instructions regarding the investigation of Mr. John Yerkes.

Very truly yours,
GILBERT HENDERSON,
Sales Manager.

ALEXANDER BOTTS
SALES PROMOTION REPRESENTATIVE
EARTHWORM TRACTOR COMPANY

BLACKSTONE HOTEL,
CHICAGO, ILLINOIS.
SATURDAY, FEBRUARY 6, 1932.

MR. GILBERT HENDERSON,
SALES MANAGER,
EARTHWORM TRACTOR COMPANY,
EARTHWORM CITY, ILLINOIS.

DEAR HENDERSON: Your letter of yesterday is here, and there is one thing about it that pleases me very much. When I wrote you that I thought you would be "delighted to know about this generous piece of rescue work," I had, all the time, a sort of a hunch that perhaps you would not be delighted at all. Your letter proves, in a very pleasing way, that I was right. It confirms an opinion I have had for a long time—to the effect that practically all my hunches are correct. And it encourages me to go ahead and act upon a very distinct hunch which has recently entered my mind.

This new hunch is a sort of vague feeling that in business, as well as elsewhere, we always get along better if we act in a large, generous and open-hearted way—looking always at the idealistic, inspirational and humanistic aspects of a problem, and refusing to stultify ourselves by a narrow preoccupation with crassly materialistic details.

Such being the case, it is obvious that I will have to follow out my original plan in this Smith matter, in spite of your orders to the contrary. But do not get the idea that I am blaming you in any way for giving me these orders. In fact, I will admit that the fault, if any, is entirely mine. My letter to you from Tylerville had to be cut short so that I could catch the train for Chicago. I did not have time to give you the details of my interview with Mr. Smith. Consequently, when you wrote me, you did not know the facts of the case. And it was perfectly natural for you to write without any real comprehension of what you were saying.

At present, fortunately, I have plenty of time. The banker whom I came to see has gone out into the suburbs somewhere for the weekend, so it will be necessary for me to remain here until he gets back on Monday or Tuesday. This will be no hardship, however, as my room here at the hotel is very comfortable, not to say luxurious. And, as I loll about comfortably on the elegant upholstery of a large armchair, I will give you an account of my visit to poor old Mr. Smith, and explain exactly why I am right in this matter and why you are wrong.

Mr. Smith's business offices had, as you know, burned down. So I called on him at his home—a handsome brick house amid pleasant surroundings in the outskirts of Tylerville. The room in which he received me was elegantly furnished and pleasing to the eye, but Mr. Smith himself was indeed a pitiable spectacle. I had remembered him, from my brief acquaintance of years ago, as a most admirable person with handsome features and a commanding presence. But all this was now changed. Before me I saw a weak and trembling creature, with a furtive expression, a nervous shiftiness of eye, and a general hangdog air of discouragement and defeat. If you had been there, Henderson, you would have felt just as sorry for him as I did. And your warm heart would have at once prompted you to do everything you could to help him.

As soon as I had introduced myself, Mr. Smith began a long and tedious discussion of his troubles. And out of his rambling remarks I was able to piece together the salient facts of the case—which are, briefly, as follows:

Mr. Smith's business is carried on through the Smith Tractor Company, a corporation in which he owns practically all the stock. The company's main asset is—or rather was—a handsome building which cost three hundred thousand dollars, and which housed the tractor salesrooms and offices. The building was mortgaged to the local building-and-loan association; the final payment of fifty thousand dollars being due next week. Besides this, the local national bank held the company's note for ten thousand—also due next week. Mr. Smith did not have the sixty thousand to make these payments—the tractor business has been a bit slow lately—but he was sure he could get an extension, or borrow the money somewhere, as the building was worth so much more than the loans, and was well insured.

And then, without warning, Mr. Smith was visited by a series of astounding misfortunes. His bookkeeper, whom he loved and trusted like his own son, fell in love with a girl who worked as assistant cashier at the building-and-loan office. They got so warm and confused about each other that they both forgot all about business. The bookkeeper forgot to mail the check that Mr. Smith gave him for the fire insurance, and his head was so full of poetry that he paid no attention to the notices that came in from the insurance company. The building-and-loan association never uttered a peep, because it was the job of the assistant cashier to check the insurance on property on which they held mortgages; and the assistant cashier, poor girl, was so busy thinking about moonlight and roses that she, naturally, couldn't keep her mind on her work. It was certainly tough on poor old Mr. Smith.

Well, the happy, though demented, pair got married a week ago Wednesday and left for Kansas City, where the bridegroom had found himself a new and better job. So that was all right for them. But the next day—Thursday—Mr. Smith's building burned down. And on Friday he was hit in the face with the news that there was no insurance, and that his company was in the hole with debts of sixty thousand, and assets, including the building lot and a few things saved from the fire, of about ten thousand.

This regrettable situation was in no way the fault of Mr. Smith. It would be unfair to claim that he should have realized the state his bookkeeper was in, and checked up on him more closely. No mere tractor man can be expected to grasp all the psychological and fiscal implications of a love affair between a bookkeeper and an assistant cashier.

The disaster was simply a case of bad luck. What happened was that the three Fates, or possibly a couple of other sinister goddesses of misfortune, sneaked up on Mr. Smith when he was not looking, and just naturally gnawed the seat right out of his pants.

And even this was not all. When the poor man realized his condition, he began rushing around to the banks, hoping to get an extension on his loans, or else to borrow enough to meet his payments. But everybody sized him up as busted beyond hope, and nobody would give him any accommodation at all. None of his friends would help him. And this final misfortune just tore the remains of his pants right off, removing in addition all that was left of his courage and morale.

At the time I visited him, he could see nothing but the dark side of things. "I am ruined," he said. "And, what is worse, I am disgraced. I am no better than a common thief. I have borrowed other people's money, and I can't pay it back. When I look at that picture up there. I feel so ashamed I almost want to shoot myself."

He pointed to an oil portrait on the wall. It showed an elderly gentleman with side whiskers.

"Who is the old bozo?" I asked.

"That is my grandfather," said Mr. Smith. "He came here to Tylerville in the '70s, and went into business. And before he died—ten years ago—he became known all over this part of the state. He was as solid and dependable

"I am ruined," he said. "When I look at that picture up there. I feel so ashamed I almost want to shoot myself."

as the United States Government. He always paid his debts. His word was his bond. And when he made a promise he always kept it. He established the reputation of the Smith family in this community. And now I have ruined that reputation. I don't see how I can stand the disgrace."

"Don't be a sap," I said. "It wasn't your fault your bookkeeper fell in love and picked out the wrong sweetie. Besides, all these debts are in the name of the Smith Tractor Company, aren't they?"

"Yes."

"And the company is incorporated?"

"Yes."

"Then you are not personally responsible for the debts at all, are you?"

"Legally I'm not. But morally I am. I own the company. So the only way I can save my reputation is to save the company. If I could save it by putting in all my personal property, I would. But I haven't enough."

"How much have you?" I asked.

"About twenty-five thousand dollars in cash and securities, and this house—it cost seventy-five thousand, but nobody seems to want to buy it."

"Have you tried to sell it?"

"I sure have. I've seen real-estate agents, and I've put advertisements in papers all over this part of the country. But there is no demand for a place like this. It's too big. People want small houses these days. The only offer so far is ten thousand, and that's probably the best I'll get. So, even if I sold out completely. I'd still owe fifteen thousand. I can't see where there is any hope for me. You had better give the dealership to Yerkes. I hear he would be glad to take it. So just forget about me. I'm nothing but a failure and a deadbeat. When I think about my grandfather—"

"For Pete's sake," I said, "forget about your grandfather. He has nothing to do with the case. He can't get you out of this hole. And you'd better forget this weak-minded, idealistic scheme of selling out your private assets to pay a lot of corporation debts. Try to be practical and optimistic, and look on the bright side of things."

"But there isn't any bright side."

"Oh, yes, there is. You've got me here to help you. And I'm one of the best little helpers in the whole United States. I'm going around to see these bankers myself. I'm going to raise the fifty thousand dollars you need. It will be in the form of a long-term loan. That will give you time to get your nerve back. You can keep on with your Earthworm Tractor dealership, and before you know it you'll have every cent paid off."

"You can't do anything with those bankers," said Mr. Smith. "I talked to them all one afternoon, and they wouldn't even listen."

"What it takes to talk to bankers, I've got," I said. "So just sit tight and don't do anything till you hear from me. And I won't sign up Mr. Yerkes without seeing you first. Good afternoon."

Leaving poor old Mr. Smith in a state of nervous collapse in his handsome residence, I went over and talked to the officers of the building-and-loan association. I told them that if they carried Mr. Smith a little longer they might get paid in full, but if they forced him into bankruptcy, he would be ruined so completely that they would get practically nothing. This seemed to me a very reasonable argument. But they replied that they had carried him long enough, that he was ruined anyway, and that they were going to collect what little they could and let it go at that. They also declined to accept any responsibility for the lapsing of the insurance, pointing out—truthfully enough, I suppose—that the insurance was primarily Mr. Smith's responsibility.

On the whole, however, they acted so dumb that I finally gave them up as a bad job, and went over to the national bank. Here I learned that nothing could be done without the authority of the president—and the president had gone off to Chicago to spend a week at a bankers' convention. This, naturally, made it necessary for me to go to Chicago myself.

I returned to the hotel, wrote a brief letter to you, and caught the late afternoon train.

I wish to point out that this trip to Chicago is not an extra expense—as insinuated in your letter—but a distinct saving to the Earthworm Company. If I had stayed in Tylerville, I would have had to waste a whole week waiting for this guy to get back. As it is, I will finish up the whole business in a few days. And the saving would have been even greater had it not been for this weekend in the suburbs.

When I called at the hotel where this bank president was supposed to be staying. I found that, instead of giving the bankers' convention his undivided attention, he had gone off to spend several days with some friends on the North Shore. It would, of course, be a tactical blunder to bother him on his vacation. So there is nothing to do but wait until his return.

And now that I have explained everything so completely and so lucidly, you will, of course, realize that I am handling this thing in exactly the right manner. However, do not get too optimistic. There is always a remote chance that this bank president may turn me down.

And it might not be a bad idea if you were to see the president of the company and tell him to arrange things so that, if we should happen to need it, he can let us have fifty thousand dollars on short notice.

Most sincerely,
ALEXANDER BOTTS.

EARTHWORM TRACTOR COMPANY
EARTHWORM CITY, ILLINOIS
OFFICE OF THE SALES MANAGER

FEBRUARY 8, 1932.

MR. ALEXANDER BOTTS,
BLACKSTONE HOTEL,
CHICAGO, ILLINOIS.

DEAR BOTTS: I regret to inform you that your handling of the Smith affair is becoming more and more unsatisfactory to this company. I have tried to be patient with you, but my patience is almost exhausted. I wanted you to go back to Tylerville and get busy at once on the job which has been assigned to you. Furthermore, I want you to handle this job in a common-sense manner—getting rid of all this mawkish and sickly sentimentality.

Kindly remember that you are working for the Earthworm Tractor Company. You are to pick out a dealer in Tylerville for this company. And in picking out a dealer, there is only one thing to consider: You must choose the man who will best serve the interests of the company.

Looking at the matter from this point of view, you should realize that Mr. Smith is completely out of the question. He has resigned; he does not want the job. Furthermore, he completely lacks the material resources with which to carry on. The Earthworm Company will not even consider advancing him any money, and, even if he gets a fifty-thousand-dollar loan from a bank, he will still be deeply in debt, and we absolutely cannot afford to depend on a man whose financial condition is so shaky.

Finally, we don't want a man who is so lacking in strength of character. Your recent letter, describing your interview with Mr. Smith, bears out this lack very clearly. Obviously, the man is a weakling. Instead of facing his troubles in a manly way, he is apparently sitting around, whining pitifully, moaning about his grandfather, and doing nothing at all to help himself. He is not the type of man we want for an Earthworm dealer. The tractor business is no place for weaklings.

I shall expect to hear, in the very near future, that you are back on the job again with all your old-time good sense and energy.

Very truly,
GILBERT HENDERSON,
Sales Manager.

ALEXANDER BOTTS
SALES PROMOTION REPRESENTATIVE
EARTHWORM TRACTOR COMPANY

BLACKSTONE HOTEL,
CHICAGO, ILLINOIS.
TUESDAY, FEBRUARY 9, 1932.

MR. GILBERT HENDERSON,
SALES MANAGER,
EARTHWORM TRACTOR COMPANY,
EARTHWORM CITY, ILLINOIS.

DEAR HENDERSON: Your letter of yesterday is received, and you will doubtless be pleased—this time I have a hunch you really will be pleased—to hear that I am starting for Tylerville this afternoon. I am leaving Chicago, partly because you seem to want me to—after all, you are the guy that hired me, and I suppose your orders are entitled to some consideration—and partly because my business here is finished up anyway. The president of the Tylerville National Bank got back to his hotel here in Chicago this morning, and I had an interview with him. On the whole, it was rather unsatisfactory.

This bank president turned out to be one of the coldest, most inhuman old money grubbers I have ever met. Besides which, he is a hypocrite—a man who attempts to justify his avariciousness and greed by spewing forth a lot of sanctimonious talk about high principles. When I suggested that he might extend Mr. Smith's note and lend him an additional fifty or sixty thousand dollars, he refused for what he called moral reasons.

"Mr. Botts," he said, "you are asking me to do something which is basically unethical. You are asking me to extend credit to a man who does not deserve credit. You seem to forget, Mr. Botts, that our whole economic structure is based on credit. And credit is based on the payment of debts. This is the first and most important principle of business. As long as people believe that a debt is something which must be paid, our business life is on a sound basis. But as soon as people begin to think that a debt is something to be wiggled out of, disaster stares us in the face."

"All of which sounds very fine," I said, "but it doesn't mean anything. What we are interested in here is not a lot of vague, general principles. We are interested in the concrete case of poor old Mr. Smith."

"Exactly. Mr. Smith owes my bank ten thousand dollars, and he owes the building-and-loan association, in which I am a stockholder, fifty thousand. He tells us he is not going to pay these debts. By defaulting in this way, Mr. Smith not only ruins his own credit, he also undermines the foundations of business in general. And any man who lends him more money under such circumstances is guilty of condoning and encouraging a very dangerous tendency."

"You've certainly got a wonderful flow of language," I said. "But didn't it ever occur to you that there is such a thing as humanity in business? And since when is it unethical to help a human being in distress? Besides, Mr. Smith is a good egg. He would pay these debts if he could. But he can't. And it isn't really his fault. You can't very well blame him because that poor sap of a bookkeeper went crazy and forgot to renew the insurance."

"I am not interested in excuses," said the bank president. "In my opinion, one of the chief troubles with the world today is the fact that there is such a large and increasing number of debtors who are using up all their energy, not in honestly attempting to pay their debts but in whining around, advancing all manner of plausible reasons why they should not pay, and even attempting to cure their troubles by borrowing more. As a matter of public duty, therefore, I must absolutely refuse to do anything that might aid or abet this vicious practice. Good afternoon, Mr. Botts."

"All right," I said, "if that's the way you feel about it, I'll not insist on casting any more pearls. Good afternoon."

With this snappy farewell, I strode jauntily from the room, concealing my natural irritation. I will frankly admit that I was pretty sore. But in spite of the fact that this filthy banker would do nothing, and in spite of the fact that you people in the Earthworm Company are so completely unsympathetic, I do not regret my efforts on behalf of Mr. Smith. It will always be a source of satisfaction to me that I did the best I could.

And now I will bow to the inevitable, and go back to Tylerville to investigate Mr. John Yerkes.

Most sincerely,
ALEXANDER BOTTS.

ALEXANDER BOTTS
SALES PROMOTION REPRESENTATIVE
EARTHWORM TRACTOR COMPANY

TYLERVILLE HOTEL,
TYLERVILLE, WISCONSIN.
WEDNESDAY EVENING, FEBRUARY 10, 1932.

MR. GILBERT HENDERSON,
SALES MANAGER,
EARTHWORM TRACTOR COMPANY,
EARTHWORM CITY, ILLINOIS.

DEAR HENDERSON: As soon as I reached Tylerville, late this afternoon, I called on Mr. Yerkes. And I will have to admit that your favorable impression of him, based on the letter which he sent you, is undoubtedly correct.

From an emotional and sentimental point of view, I hate to see Mr. Smith getting washed down the sewer. But you are right in saying that this affair must be decided from a strictly business point of view. We must get the man who will best serve the interests of the Earthworm Tractor Company. And Mr. Yerkes seems to have the three qualifications which you so logically set forth in your letter as being essential. He is very eager to get the job, and will undoubtedly push our business here with great vigor and efficiency. He has ample capital. And finally, he has all the strength of character, optimism and virility which poor old Mr. Smith so conspicuously lacks. But the thing that impressed me the most about Mr. Yerkes was the truly noble way in which he agreed with my opinion of that mephitic reptile, the president of the Tylerville National Bank. During my conversation with Mr. Yerkes I had told him some of the idiotic remarks the bank president had made to me in Chicago. And I had also repeated, with justifiable pride, a few of the snappy replies I had handed the old crab.

"You are a man after my own heart!" Mr. Yerkes said. "Only I'm afraid you didn't give it to him hot enough. You should have heard me the last time I told him what I thought of him."

"So you have had trouble with him too?"

"Plenty. You wouldn't believe it, but he once tried to get me to pay him a debt I didn't owe."

"How could he expect to get by with anything as raw as that?"

"Oh, it's a long story. A few years back I had a run of bad luck—lost a lot of money in Florida real estate and other things—and the first thing I knew my implement business here had gone busted on me. I had to let it go through bankruptcy. The creditors got twenty cents on the dollar—and they were lucky at that. Fortunately, the business was incorporated, so they couldn't touch my private assets. I had plenty of funds to get started again, and in about three years I was more prosperous than ever. It was a good comeback, and I don't mind saying I'm proud of the way I managed things. But it certainly made this bank president sore."

"You mean he hated to see anybody besides himself making any money?"

"Yes. And he had a fool idea that I had done him dirt in some way. You see, when my company went busted, his bank was the principal creditor. It like to drove him crazy, getting only twenty cents on the dollar. And when I was back on my feet again, he wrote me a nasty letter asking me to pay the balance, with interest. Can you tie that? He actually expected interest too."

"And you wrote him a letter telling him where to get off?"

"Wrote him a letter, nothing. I went down there personally. I took the letter, and I waved it in his face, and I said, 'Did you write this?' 'Yes,' he said. So I says, 'What's the idea? Don't you know you got no legal claim on me at all?' And he says, 'That's true. But now you've got money, you ought to pay your back debts just the same. It's a moral obligation.'"

"He's always drooling out that pious applesauce about morals," I said. "What did you tell him next?"

"Believe me, I told him plenty. I just poured it onto him. 'The laws of this great country,' I says, 'tell me I don't have to pay you. But you think you know better. Is that showing a proper respect for the law? Is that good citizenship? Is that what you call good morals?'"

"A neat crack," I said. "What answer did he make?"

"Well, sir, he didn't have a word to say. I had him, and he knew it. So I went right on. I said, 'The law is just, and enlightened, and based on human values. It provides the corporation form of doing business so a man's commercial losses can't ruin him personally. It provides bankruptcy proceedings, so a man can get rid of his debts and make a fresh start. Even in international finance, it is now admitted that government debts must be adjusted on the basis of capacity to pay. And here you are talking as if we lived in the dark ages. If a man once gets in debt, you want to grind

"I took the letter, and I waved it in his face, and I said, 'Did you write this?'"

him down and keep him down forever. I bet you would even like to have debtors' prisons again.'"

"I bet he would, at that," I said.

"So then," Mr. Yerkes continued, "I told him he had no cause for complaint anyway. I said, 'You bankers charge so much interest on all your loans that you have more than enough to cover the ones that go bad. If you didn't have a few losses, your profits would be all out of reason. You've got more money than I have anyway. So why should I, for no reason at all, turn over a lot of my hard-earned cash to a guy like you that is already lousy with wealth? Answer me that!'"

"And did he answer you?" I asked.

"No. He just sat there. So I bawled him out a bit more. And finally I took the letter he had sent me, and I threw it in his face. 'You can take that,' I says, 'and you know what you can do with it.' Then I walked out."

"Swell!" I said. "I sure wish I had been there to hear you."

I was so pleased with Mr. Yerkes's account of the way he had showed up this bank president for the slimy bum that he is that I wanted to sign a dealer's contract with him right away. But I had promised Mr. Smith I would see him before I placed the contract elsewhere. So I told Mr. Yerkes that I would give him a definite answer about the dealership tomorrow

"And finally I took the letter he had sent me, and I threw it in his face."

afternoon. We then shook hands in a most cordial fashion, and I came back to the hotel, arriving just in time for supper.

I will see Mr. Smith tomorrow morning, and I shall hope to get everything fixed up with Mr. Yerkes in the afternoon, so that I may leave for the next job before night.

Most sincerely,

ALEXANDER BOTTS.

ALEXANDER BOTTS
SALES PROMOTION REPRESENTATIVE
EARTHWORM TRACTOR COMPANY

TYLERVILLE HOTEL,
TYLERVILLE, WISCONSIN.
THURSDAY NOON, FEBRUARY 11, 1932.

MR. GILBERT HENDERSON,
SALES MANAGER,
EARTHWORM TRACTOR COMPANY,
EARTHWORM CITY, ILLINOIS.

DEAR HENDERSON: When I called at Mr. Smith's handsome residence this morning, I found the whole place locked up and apparently deserted. I went to the house next door, where I found an elderly lady who told me a most pitiful tale.

"Poor Mr. Smith!" she said. "His house has been sold, and he and Mrs. Smith and the two children have moved away. It's a very sad case. They are such fine people. They have been our neighbors for years. And now they are being driven out of their home—which seems so cruel and unjust. I can't see why—just because Mr. Smith failed in business—they should take away his home and throw the whole family penniless into the streets."

"I don't understand it myself," I said. "I know something of Mr. Smith's affairs, and I understood that Mr. Smith's house would be safe from his creditors. But, evidently, I was mistaken. How does Mr. Smith seem to be standing the shock?"

"He is putting up a very brave appearance. But, inwardly, he must be in despair. Really, it's pitiful."

"Yes," I agreed, "it is. When did the Smiths move? And where did they go?"

"They left two days ago," said the lady, "and I understand they have rented a miserable little shack over on the other side of the railroad. And even that isn't the worst of it. Apparently, Mr. Smith has lost every cent of his money, and, in order to eke out a bare living, he will have to humiliate himself by doing manual labor."

"What sort of manual labor?"

"It's most extraordinary. He has been going around to all his old friends, including myself, and asking for the job of running our furnaces in the winter and taking care of our lawns in the summer. We have all been dreadfully embarrassed at the idea of employing an old friend for such menial tasks. But as long as he needed the work and asked for it, we have agreed to give it to him. It's a wretched situation."

"It surely is," I agreed. "Can you tell me where I can find this unfortunate gentleman?"

She gave me the address. I thanked her and departed. After twenty minutes' walk, I found myself at Mr. Smith's new residence. It was in a rather cheap section of town, not far from the freight yards. But it did not quite deserve the description of "a miserable little shack." It was a small, inexpensive, but rather cute bungalow. I walked up to the door with some hesitation. I rather dreaded an interview with this miserable man. But when Mr. Smith answered my knock. I saw at once that his former neighbor had been right when she said he was putting up a brave appearance. As a matter of fact, his appearance was so remarkably brave that I could hardly believe I was talking to the same man who, only last week, had been so downcast, so miserable, and so filled with despair.

"Good morning, Mr. Botts," he said, shaking hands enthusiastically. "Come in." His manner was cordial and friendly, and he seemed to be fairly bubbling over with enthusiasm and good nature. He led me into a small and rather cheaply furnished living room.

"I am glad," I said, "to see you looking so cheerful, Mr. Smith."

"Why shouldn't I be cheerful?" he answered. "I have just had a wonderful piece of good luck. You remember what a ghastly mess I was in when you came to see me last week?"

"I certainly do."

"At that time," Mr. Smith went on, "I didn't think I could ever get out of it. And then, the day after your visit, the miracle happened."

"What happened?" I asked.

"A wealthy gentleman in Milwaukee saw one of the newspaper advertisements in which I offered my house for sale. He wanted a place in this part of the country. He was a fast worker. He came. He saw. He bought. And, instead of ten thousand, which was the best previous offer, this bozo paid twenty-five thousand for the whole place, including all the furnishings, except my grandfather's portrait and a few personal belongings. As soon as the deal went through, I sold all my other assets and paid up every dollar owed by the Smith Tractor Company. I tell you. Mr. Botts, it was wonderful. Never in all my life have I been so happy."

"But it leaves you absolutely flat, doesn't it?"

"No. After everything was fixed up, I found I had $37.42 left over, so I am that much ahead of the game. Besides, I still have my grandfather's portrait. So everything is fine."

"Just the same," I said, "I should think it would be rather distressing to you to have to move out of your lovely home into this place. How does the rest of your family feel about it?"

"They are behaving swell," said Mr. Smith, with a somewhat self-satisfied smile. "For the first time in years, my wife is doing all the cooking and housework, and she claims she likes it. My two sons are continuing in school, and entirely on their own initiative, they have decided to help the family finances by starting up a magazine-subscription business. You don't know how good this makes me feel. I never realized, until just lately, what a fine family I have."

"Is it true," I asked, "that you are planning to take care of lawns and furnaces for a living?"

"Exactly," said Mr. Smith. "The work doesn't pay very much, and eventually, of course, I may get into something better. But for the present it will do very well. The rent of this house here is very low and we'll have enough to live on, simply, but very comfortably. So, you see, there's nothing to regret about this whole business, and there is everything to be thankful for."

"Well, maybe so," I said. "It's true that you might be a whole lot worse off. But I can't understand why you should be so thankful about things."

"Don't be a boob," said Mr. Smith. "Who wouldn't be thankful in my position? I've paid all my debts. I don't have to worry anymore. I don't have to go whining and slinking around. When I go downtown, I can hold up my head and strut around as proud as I want. And when I meet any of these birds who made the mistake of thinking I was going to be a low-down bankrupt and defaulter, I can laugh right in their faces. I can thumb my nose at them."

"I hope," I said, "that when you meet the president of the national bank you will use both hands and plenty of wiggling of the fingers."

"But I don't want to thumb my nose at him. He has acted too decent."

"What?" I said. "That old, frozen codfish?"

"Oh, yes. He's not really so bad as you think. He came around last night, after he got back from Chicago, and told me that I had worked this thing out in exactly the right way. He said he was very much pleased by the whole business."

"Of course, he was pleased," I said. "There's nothing gives that guy greater joy than seeing a man sold out and left flat. And the flatter the better."

"No," said Mr. Smith, "you're wrong about that. The old boy was actually sorry I'd lost everything, and he said that if I wanted to go on with the tractor business, he would finance me."

"In what way?"

"We would form a new company. He would supply all the capital and take half the stock. I would run the business and gel the other half of the stock, plus five thousand dollars yearly as salary. It seems to me that's a pretty liberal offer."

"Yes," I admitted, "it's a swell offer. But I can hardly believe he made it. When I talked to him, he wouldn't even risk five cents on you."

"But that was when he thought I wasn't going to pay my debts. Now he knows I have paid them, and he also knows I put in a lot of my personal assets that I was not legally required to contribute. At present, he says, my credit is just as good as anybody else's cash. And he says he is not making this offer with the idea of helping me. He claims it is a purely business affair. He is always glad to invest his money in a good, safe business which promises to make him a handsome profit."

"Probably that's it," I said. "He's perfectly selfish, after all. But from your point of view, it looks to me like a good proposition. What did you tell him?"

"I thanked him and told him I could not accept."

"Why not?"

"In the first place, I very much doubted if you people would renew my contract. You've practically decided to give it to Mr. Yerkes, haven't you?"

"I had intended to," I said, "and Yerkes is very anxious to have it. But I haven't made any definite promises."

"You'd better give it to him," said Mr. Smith, "because, on the whole, I think I'd rather not take it, even if it was offered to me. I've had so much worry lately that I need a vacation. I think I'll stay out of any sort of executive business job—for a while, at least. This taking care of furnaces and lawns will mean a certain amount of physical work, but it will be a rest for my mind. I'm actually looking forward to it with pleasure. So don't worry about me, Mr. Botts. Now that I'm out of debt, I'm really on top of the world. I don't have to ask any favors from anybody. And I can look my grandfather's portrait right in the eye and tell myself that I am just as good a man as he was."

"Well, if that's the way you feel about it, Mr. Smith," I said, "there's nothing much more I can say, except that I am very glad indeed to see you are taking this thing in such a fine spirit. Goodbye and good luck to you."

"Goodbye," said Mr. Smith, "and give my best wishes to Mr. Yerkes."

I walked back to the hotel, and I have been spending the rest of the morning writing this report. As it is now lunchtime, I will close. This afternoon I will see Mr. Yerkes, and get this business finished up.

Very truly,
ALEXANDER BOTTS.

TELEGRAM
EARTHWORM CITY ILL 10 AM FEB 12 1932
ALEXANDER BOTTS
TYLERVILLE HOTEL
TYLERVILLE WIS

HOLD UP NEGOTIATIONS WITH YERKES STOP AFTER
READING YOUR TWO LATEST REPORTS I AM INCLINED TO
THINK WE OUGHT TO URGE SMITH TO TAKE CONTRACT

GILBERT HENDERSON

———

ALEXANDER BOTTS
SALES PROMOTION REPRESENTATIVE
EARTHWORM TRACTOR COMPANY

TYLERVILLE HOTEL,
TYLERVILLE, WISCONSIN.
FRIDAY NOON, FEBRUARY 12, 1932.

MR. GILBERT HENDERSON,
SALES MANAGER,
EARTHWORM TRACTOR COMPANY,
EARTHWORM CITY, ILLINOIS.

DEAR HENDERSON: Your telegram has just arrived, and it is too late. The contract has already been signed. But when I explain all the circumstances, I am sure you will agree that I have acted wisely.

Yesterday noon, after I had written you about my visit to Mr. Smith, I decided it might be well, before proceeding further in this matter, thoroughly to weigh and consider the various new factors which had entered the ease. Accordingly, instead of calling on Mr. Yerkes at once, I retired to my room and gave myself up to a prolonged period of deep meditation and thought. And it gives me great pleasure to report that I finally arrived at a singularly brilliant idea which cuts through all the fogs and uncertainties of this dealership question, and makes the whole proposition as clear as day. This idea is so unusual that not one man in ten thousand would have thought of it. But it is so simple, when once it has been stated, that even a guy like you will understand it and recognize its truth and validity.

In evolving this idea I started with your proposition that this is strictly a business affair, in which sentiment has no place. In picking a dealer here, we must choose the man who will best serve the interests of the Earthworm Tractor Company. Having adopted this principle, I took up your assertion that "the tractor business is no place for weaklings." And, after a long process of careful analysis and logical reasoning, I reached the conclusion that this second idea was correct.

I then compared Mr. Yerkes and Mr. Smith. And it was at once evident that the magnificent way Mr. Yerkes talked up to the bank president indicated a remarkable amount of strength and virility. On the other hand, the pathetic manner in which Mr. Smith talked of his troubles and harped on his grandfather indicated nothing but weakness and lack of force.

An ordinary man would have decided at once in favor of Mr. Yerkes. But I did not. I kept on doggedly and persistently turning the matter over in my mind—thinking, analyzing, considering. And at last, after four hours and twenty minutes of the most intense mental activity, the big idea burst upon me. I realized, with a sudden flash of inspiration, that the way these men talk is of no more importance to us than a pinch of bug dust—the only thing that matters is the way they act. Looking at it from this point of view, it was clear that Mr. Smith, in spite of his sentimental gabblings, had really done something when he paid off his debts. And this led me right on to the astounding conclusion that the real weak sister was the virile and impressive talker who had been perfectly satisfied to let his creditors get along with twenty cents on the dollar.

I was, naturally, delighted that my original hunch in favor of Mr. Smith was thus confirmed by purely intellectual reasoning. And I have been surprised and amused to learn, from your telegram, that you—probably by accident—have also hit upon the correct point of view.

As soon as I reached my inspired decision, I dropped the role of cloistered scholar and became the man of action. I rushed over to Mr. Smith's house. I told him that the Earthworm Tractor Company could not get along without him. I brushed aside his plea that he wanted a vacation among the furnaces and lawn mowers. And I am overjoyed to announce that, after considerable argument, he gave in and signed the contract, which I enclose herewith.

Very truly,
ALEXANDER BOTTS.

WORKING ON THE LEVEE

ILLUSTRATED BY TONY SARG

ALEXANDER BOTTS
SALES PROMOTION REPRESENTATIVE
EARTHWORM TRACTOR COMPANY

MEMPHIS, TENNESSEE.
SATURDAY, AUGUST 20, 1932.

MR. GILBERT HENDERSON,
SALES MANAGER,
EARTHWORM TRACTOR COMPANY,
EARTHWORM CITY, ILLINOIS.

DEAR HENDERSON: I arrived in Memphis this morning, called at once on our local dealer, Mr. Fitz William, and ran into a piece of news which so filled my bosom with joy that it like to shot all the buttons off my vest.

Mr. Fitz William says that the Federal flood-control people are on the point of spending millions of dollars for the building of many miles of levees in the Yazoo-delta region along the Mississippi River to the south of here. In other words, while the hot sands of the desert of depression are drifting drearily over most of the country, there is, in this region, an amazing oasis of prosperity. And this oasis ought to be a swell place to sell tractors.

At the moment, the best prospect seems to be a local contractor by the name of Jim Slanker, who has just been awarded a two-million-yard levee-building contract. He is said to be in the market for twenty-five sixty-horsepower tractors and ten elevating graders, besides a lot of shovels, drag lines and other equipment.

Mr. Slanker lives here in Memphis, but he has gone down into the Yazoo delta, south of Greenville, to a small place called South Gumbo, Mississippi, which is the site of the levee he is to build. Mr. Fitz William, at my suggestion, is shipping a sixty-horsepower Earthworm tractor and an Earthworm elevating grader to South Gumbo. And he and I are starting down ourselves this afternoon to put on a demonstration.

This begins to look like the good old days. We got a chance to sell twenty-five tractors and ten graders! You will agree with me that it sounds almost like a fairy tale.

My address for the next few days will be South Gumbo, Mississippi.

Hastily but joyfully,
ALEXANDER BOTTS.

NIGHT LETTER TELEGRAM
EARTHWORM CITY ILL AUG 22 1932
ALEXANDER BOTTS
SOUTH GUMBO MISS

DELIGHTED WITH YOUR LETTER STOP IN THE PRESENT HARD TIMES A SALE OF TWENTY FIVE TRACTORS WOULD MAKE ALL THE DIFFERENCE BETWEEN CONTINUING FACTORY PRODUCTION AT CURRENT REDUCED RATE OR LAYING OFF MORE MEN AND PARTIALLY SHUTTING DOWN STOP IN VIEW OF THE GREAT IMPORTANCE OF THIS DEAL I AM PLANNING TO COME DOWN TO MISSISSIPPI IN A FEW DAYS TO TAKE CHARGE PERSONALLY

GILBERT HENDERSON

———

ALEXANDER BOTTS
SALES PROMOTION REPRESENTATIVE
EARTHWORM TRACTOR COMPANY

SOUTH GUMBO, MISSISSIPPI.
TUESDAY EVENING, AUGUST 23, 1932.

MR. GILBERT HENDERSON,
SALES MANAGER,
EARTHWORM TRACTOR COMPANY,
EARTHWORM CITY, ILLINOIS.

DEAR HENDERSON: Your telegram has come, and it is my painful duty to inform you—with all due respect for your high character and excellent intentions—that it would, on the whole, be much better, in my opinion, for you to stay up there in the office where you belong, and not try to come down here and start monkeying with a situation that you don't know anything about. In case you might be inclined to doubt the wisdom of my advice, I will give you a brief account of my activities for the past few days, explaining all the trouble I have had, and clearly setting forth the reasons why the present unfortunate situation can best be handled by me without any outside interference.

Mr. Fitz William, our Memphis dealer, and I arrived here at South Gumbo day before yesterday. The town is nothing but a railroad station and two or three houses, set in the midst of the desolate swamps and canebrakes along the Mississippi River. We found Mr. Jim Slanker at his newly established construction camp not far from the station. He is a great big bozo, more than six feet tall, heavily built, and with a voice like the whistle on the factory at Earthworm City. His chin carries a heard of several days' growth, and his hip is decorated with a large revolver in a holster.

The camp has just been started, but already there is quite a force of men at work—engineers and surveyors going over the site of the levee, mechanics setting up equipment, and a gang of workmen putting up bunk houses and tool sheds. When we arrived, old Jim was bossing around a bunch of these workmen in a way that was slightly louder and more hard-boiled than a regular-army first sergeant.

Waiting for a lull in the proceedings, we introduced ourselves. Old Jim at once said he would be very glad to see our tractor, and invited us to stay at his camp. Like most of these tough babies, he is hospitable and bighearted. And we would have got along fine except for one thing.

This one fly in the ointment was Mr. Fitz William, our dealer from Memphis. Before he got through, this guy pretty near ruined all our chances. It wasn't that he deliberately tried to ball things up, or that he did any one thing that was so bad.

The whole trouble was that he is a white-collar man from the city, and when he found himself down in the swamps with a bunch of hard-boiled dirt movers, he was lost. Everything he did was just a little bit wrong. And all the time that I was attempting—by means of my natural tact and intelligence—to get us in strong with old Jim and his organization, this guy was gradually but steadily queering everything.

In the first place, he insisted on telling everybody exactly who he was, which wouldn't have been so bad except that he gave them his full name, which turns out to be Edwin Reginald Fitz William. Can you tie that?

Then he started in wearing the damnedest clothes—which some store in Memphis sold him as the proper thing for roughing it in the great out-doors. The outfit included a cute little Norfolk jacket, golf knickerbockers, shiny riding boots, a white shirt with a high stiff collar, and a Tyrolean hat with a little feather in it. As far as his costume was concerned, this feather was the last straw—or, perhaps, quill.

But even so, he might have got by if he hadn't talked so much. The tractor and the grader were delayed somewhere in shipment, and didn't arrive until late this afternoon. For two days we had nothing much to do. And Mr. Edwin Reginald Fitz William insisted on tagging around after old Jim Slanker prattling endlessly about the advantages of Earthworm tractors and graders. I kept telling him that Jim was busy and didn't want to be bothered. I pointed out that the demonstration we were going to put on would be far more convincing than any mere line of talk. But he would not listen to me.

Old Jim held onto himself for a long time. Evidently, he did not want to get rough with a man who was a guest at his camp. But his irritation kept increasing. Twice on Monday afternoon he quietly advised Edwin to shut up. Early this morning—Tuesday—he told him fairly loud that he had better shut his face. And just before lunch, he not only yelled at him in a very loud voice and told him to shut his face but also pushed him out of the office and slammed the door.

This kept the poor sap quiet for several hours. But when the tractor and the grader finally arrived on a freight car, late this afternoon, he rushed up to Jim's office to announce the glad news. And, in spite of the fact that Jim

He yelled at him in a loud voice and pushed him out of the office.

was busy with one of his engineers, he started a long oration about all the wonderful things the tractor was going to do tomorrow.

So Jim just took him by the collar, stuck the end of his gun into his back, marched him down to the station, put him on the evening train for Memphis, and told him that, if he ever showed his nose around here again, he would shoot him just like a rabbit. And as the train pulled out, old Jim turned to me and said that I could stick around as long as I behaved myself.

"But," he added, "if you ever bring down any more of these babbling white-collar friends of yours, I will throw you out, just like Fitz William."

So that was that. And now I think you can appreciate my consternation when I received—a few minutes after Mr. Fitz William's departure—your telegram announcing that you were coming down here. And I think you will understand my insistence that you stay right where you are. The presence, for only two days, of one white-collar man has almost ruined my chances here. The arrival of another would cause the entire deal to blow up with a loud report.

And even if you were going to be a help instead of a hindrance, your presence would be totally unnecessary. I am perfectly competent myself. Considering the circumstances, I have accomplished a whole lot already. Of course, up to this time, all my efforts to get in strong around here have been neutralized by the egregious asininities of the sap from Memphis. On the other hand, however, I have not aroused any active antagonism. And now that the big annoyance has gone home, my pleasing personality will begin to get in its work. The demonstration of the tractor and grader, which I expect to start tomorrow morning, will naturally be a tremendous help. And before long I shall doubtless have old Jim Slanker just eating out of my hand.

But even this is not all. In addition to my direct approach to Mr. Jim Slanker, I have already started a cautious reconnaissance with a view to making a sort of flank attack.

This plan is based on my previous experience with a great many other hard-boiled bozos. I have found that these people always have a soft spot somewhere. The tougher they are, the more apt they are to have someone closely associated with them upon whom they lean for advice and counsel. Perhaps you remember old man Higginson in Omaha. Several of our bright salesmen had been chasing him for years—and they never got anywhere, because he wouldn't buy anything without the OK of his chief mechanic, who was such a shy, retiring lad that none of the boys knew he existed until I came along and put over a sale by working on him instead of on the old man.

Of course, it is not always the chief mechanic. Sometimes you have to see the guy's wife. Sometimes he has a woman secretary who is hired to run the office, but who also runs pretty much the whole business, including the boss. Sometimes, of course, the boss makes all his own decisions without consulting anybody. But in the case of Mr. Jim Slanker, I have a very strong hunch—and my hunches are almost always right—that there is somebody in his organization who has the inside track and can make or break this sale.

As soon as I discover this power behind the throne, I will start working on him, or her, as well as on the old man, and the sale will go through as easy as a toenail through a sock. There isn't even any competition from rival tractor companies; up to this time, no other salesmen have penetrated this far into the swamps. So all you have to do, Henderson, is stop worrying and leave everything to

Your thoroughly competent representative,
ALEXANDER BOTTS.

P.S. I hope you won't be offended at my telling you so positively not to come down here. My advice is based on a sincere desire to further the best interests of the Earthworm Tractor Company. And all my remarks have been made in a spirit of the greatest possible friendliness. No one, Henderson, has a greater admiration for your sterling qualities than I. And there is no one who appreciates more fully your brilliancy and efficiency as the executive head of the sales department of the Earthworm Tractor Company.

But you have got to admit that your proper place is at the factory. Your natural sphere of operations is the office, not the field. You spend so much time in a swivel chair that you are just a pants polisher, a white-collar city slicker and, as such, totally lacking in that adaptability which is so necessary for a man who expects to succeed in the field.

I, on the other hand, am the ideal field man—the quintessence of adaptability. When I am in the swamps, I am just like a little duck. When I am in the Rockies, I am like a mountain goat. And when I am in the Arizona Desert, you could not tell the difference between me and a Gila monster. Furthermore, when it comes to handling tough guys like Jim Slanker I am there. So you had better just stay home, Henderson, and leave this job down here to a bozo that understands it.

NIGHT LETTER TELEGRAM
EARTHWORM CITY ILL AUG 25 1932
ALEXANDER BOTTS
SOUTH GUMBO MISS

YOUR LETTER RECEIVED STOP DON'T WORRY ABOUT
MY BEING OFFENDED STOP I KNOW EXACTLY HOW YOU
FEEL STOP WHEN I WAS A SALESMAN MYSELF IN THE
OLD DAYS I USED TO HAVE THE SAME DISTRUST OF
THE MAN WHO WAS THEN SALES MANAGER STOP AND
DONT WORRY ABOUT MY SPOILING THINGS IF I COME
DOWN STOP I DONT CLAIM TO BE SO ADAPTABLE THAT
PEOPLE WOULD MISTAKE ME FOR A DUCK GOAT OR
GILA MONSTER BUT BEFORE I WAS SALES MANAGER I
SPENT MANY YEARS IN THE FIELD AND SUCCESSFULLY
HANDLED MORE TOUGH GUYS THAN YOU HAVE EVER
SEEN STOP HOWEVER I AM VERY BUSY HERE AND IF YOU
ARE ABSOLUTELY SURE YOU CAN SWING THIS DEAL
I MAY NOT COME DOWN STOP WILL LET YOU KNOW
DEFINITELY IN A FEW DAYS

GILBERT HENDERSON

ALEXANDER BOTTS
SALES PROMOTION REPRESENTATIVE
EARTHWORM TRACTOR COMPANY

METHODIST EPISCOPAL HOSPITAL,
VICKSBURG, MISSISSIPPI.
SATURDAY, AUGUST 27, 1932.

MR. GILBERT HENDERSON,
SALES MANAGER,
EARTHWORM TRACTOR COMPANY,
EARTHWORM CITY, ILLINOIS.

DEAR HENDERSON: Your telegram was forwarded from South
Gumbo, and has reached me down here at the hospital. I wanted to write

to you before, but I have not been feeling so good. Today, however, for the first time, I am pretty comfortable. So I will now give you all the news.

On Wednesday morning I put on my scheduled tractor and grader demonstration, using one of Mr. Slanker's mechanics as tractor operator. Everything went swell. As we rolled along through the canebrake, plowing up the tough gumbo soil and casting it onto the site of the new levee, I could see that both Mr. Slanker and the operator were deeply impressed. Neither one of them had ever before seen our new and improved models at work.

The real excitement, however, came during the noon hour, in the form of an incident which proves that I was right when I told you in a former letter that this is a tough country, and no place for a city slicker.

At exactly twelve o'clock I stopped the tractor, and Jim Slanker and the operator and I sat down to eat our lunch beside a big two-yard steam shovel, which had been opening up a drainage ditch, and which had just stopped work for the noon hour. We had hardly seated ourselves before there suddenly appeared, on the other side of the steam shovel, four men from the gang who had been working at the camp. They were drunk. And they were in an ugly mood. Two of them had knives. The others had pick handles. Advancing in a compact group, they began a highly threatening and somewhat incoherent demand for higher wages.

Now, it is a basic principle down here in the swamps that a contractor must never show any weakness or stand for any foolishness from his workmen. If he does, his prestige is gone and he is done for. It was obviously up to Jim Slanker to do something. And he did. He was the only one of us who was armed, and he promptly pulled out his gun and ordered the four men to go back where they belonged.

Unfortunately, however, they were too drunk to know what they were doing. They kept slowly advancing upon us, coming around from behind the huge dipper of the shovel, which was resting on the ground. Again Jim shouted a warning. They paid no attention. It seemed only a question of seconds before he would start firing.

At this critical juncture—and I may say that of all the junctures I have experienced, this was by all odds the most critical—I suddenly got an idea. In less than a second I figured out a comprehensive plan for the handling of this dangerous situation. And almost in the twinkling of a gnat's eyelash, I not only put this plan into action but carried it through to a triumphant conclusion. Truth compels me to admit that immediately afterward I blundered into an idiotic and disgusting anticlimax. But that does not alter the fact that the main part of the show was a real honey—

one of the cleverest and most sensational performances which has ever been pulled in the entire history of the dirt-moving business.

And the whole thing was so charmingly simple that it was over in less than one-tenth of the time it takes to tell it. I sprang to my feet. I cast from me the remains of the ham sandwich on which I had been gnawing. Quickly but silently I climbed into the cab of the shovel, where a glance at the gauge showed me that the steam pressure was OK. Standing at the controls, I watched the four menacing workmen. They were all bunched up together. They scowled. They threatened. They advanced. They moved around in front of the dipper.

And then I launched my attack. With a single flip of the wrist I sent the huge dipper lunging forward, neatly scooping them up; and then, swinging the dipper, with them in it, high in the air, I brought it to rest at a point about forty feet above the surface of the ground.

I sent the huge dipper lunging forward, neatly scooping them up.

Jim Slanker and the operator were so astounded by this sudden change in the situation that they seemed unable to comprehend what had happened. So I called out to them, assuring them that everything was now all right.

"You don't have to worry anymore," I said. "I have got those babies where they can't hurt anybody. All you have to do is talk to them in a fatherly manner, and I think they will see the error of their ways."

At this point, four bewildered faces appeared, looking over the rim of the dipper. And never have I seen people change so completely in such a short time. My little gesture with the steam shovel had produced a psychological as well as a physical effect. The suddenness and unexpectedness of the thing had been such a shock to the four victims that all the fight was gone out of them. They were helpless, and they knew it. The dipper was too high for them to jump out, and they were in such a state of nervous collapse that they didn't dare try to crawl down along the boom. In other words, they were completely humbled, ready to beg for forgiveness and do anything that Jim Slanker asked them. My modest effort had been crowned with complete success.

As I stepped down from the steam shovel, I was so filled with pride and admiration for myself that I fear I did not pay sufficient attention to where I was stepping. And this, as it turned out, was just too bad. My foot slipped, and I fell heavily to the ground. The drop was not more than three feet, but my foot got doubled under me in some way, and as I landed I sustained an injury which later turned out to be a broken ankle. It was a most distressing and idiotic ending for my great and heroic exploit.

Jim Slanker acted splendidly. Leaving the operator to deal with the four workmen, he rushed me in his car all the way down here to the hospital at Vicksburg, and he did not return to South Gumbo until the broken bone had been set and I had been put to bed.

There is not much more to tell. The doctor says the fracture is a simple one, and will heal up as good as new. By this time I am comfortable enough, but I will have to stay in bed a week or two longer. This is, naturally, very annoying, as I ought to be back at South Gumbo.

However, I have a feeling that everything will come out all right, in spite of my absence. Yesterday I got a letter from Jim Slanker saying that his operator is continuing the demonstration. Mr. Slanker wants to run the tractor for at least a week longer in order to find out whether it can keep on moving dirt at the same rate at which it started out.

Mr. Slanker says that his operator finally released the four men after

WORKING ON THE LEVEE

giving them a terrific bawling out. He says they were so cowed that he is sure they will never make any trouble again.

Incidentally, Mr. Slanker is highly grateful for my part in the affair. "If it had not been for you," he writes, "I would have had to shoot those bums. And shooting is always messy; it makes a lot of trouble, excites the rest of the help, and causes a lot of extra work digging graves. I was too dumb to think of anything else. But you have brains. And any guy who can go up against four of these critters, drunk and waving knives and pick handles, and, by using a steam shovel, handle them gently and almost tenderly, yet with complete firmness—any guy that can do that is a real man. So I take off my hat to you, Mr. Botts."

In closing, I may say that I consider this a very touching tribute. I think I will save Mr. Slanker's letter for my grandchildren.

No more at present.

As ever,
ALEXANDER BOTTS.

TELEGRAM
EARTHWORM CITY ILL AUG 28 1932
ALEXANDER BOTTS
METHODIST EPISCOPAL HOSPITAL
VICKSBURG MISS

SORRY TO HEAR OF ACCIDENT AND HOPE EVERYTHING IS GOING WELL STOP WISH YOU HAD NOTIFIED ME SOONER STOP I AM LEAVING AT ONCE FOR SOUTH GUMBO TO TAKE CHARGE OF DEMONSTRATION STOP WILL GO BY WAY OF MEMPHIS AND SEE FITZ WILLIAM STOP PLEASE WIRE ME THERE ANY INFORMATION YOU FEEL MIGHT BE HELPFUL STOP IN A FORMER LETTER YOU MENTIONED THE POSSIBILITY THAT THERE MIGHT BE SOMEONE IN SLANKERS ORGANIZATION WHO WOULD HAVE DECISIVE INFLUENCE WITH HIM IN THIS DEAL STOP PLEASE LET ME KNOW WHAT YOU FOUND OUT AND ADVISE WHAT YOU THINK WOULD BE MY BEST PLAN OF ACTION

GILBERT HENDERSON

TELEGRAM
VICKSBURG MISS AUG 28 1932
GILBERT HENDERSON
CARE EDWIN REGINALD FITZ WILLIAM
MEMPHIS TENN

I GOT FULL DOPE ON SLANKERS ORGANIZATION BEFORE
I LEFT SOUTH GUMBO STOP THE PERSON FOR YOU TO
SEE IS MISS FLORA HIGGINS MR SLANKERS SECRETARY
WHO IS IN CHARGE OF HIS MAIN OFFICE IN MEMPHIS
STOP GET IN TOUCH WITH HER AT ONCE AND STAY
WITH HER UNTIL YOU HAVE HER COMPLETELY SOLD ON
EARTHWORM TRACTORS AND GRADERS

ALEXANDER BOTTS

TELEGRAM
MEMPHIS TENN SEPT 1 1932
ALEXANDER BOTTS
METHODIST EPISCOPAL HOSPITAL
VICKSBURG MISS

ARRIVED IN MEMPHIS DAY BEFORE YESTERDAY STOP
HAVE SPENT MOST OF MY TIME SINCE TRYING TO SELL
MISS FLORA HIGGINS ON THE IDEA OF EARTHWORM
EQUIPMENT BUT WITHOUT ANY TANGIBLE RESULTS
STOP I AM BEGINNING TO HAVE VERY GRAVE DOUBTS
AS TO THE WISDOM OF THIS COURSE STOP ARE YOU
ABSOLUTELY SURE ABOUT THIS WOMANS INFLUENCE IN
DECIDING ON TRACTOR PURCHASE STOP DO YOU STILL
ADVISE MY REMAINING HERE AND FOLLOWING THIS
LINE STOP WIRE REPLY

GILBERT HENDERSON

TELEGRAM
VICKSBURG MISS SEPT 1 1932
GILBERT HENDERSON
CARE EDWIN REGINALD FITZ WILLIAM
MEMPHIS TENN

YOUR TELEGRAM RECEIVED AND I HASTEN TO ASSURE
YOU THAT I HAVE ABSOLUTELY NO DOUBTS WHATEVER
REGARDING MISS HIGGINS INFLUENCE ON SLANKER STOP
I CANNOT HOWEVER GIVE DEFINITE ADVICE AS TO YOUR
COURSE OF ACTION WITHOUT KNOWING EXACTLY WHAT
YOU HAVE ACCOMPLISHED SO FAR STOP PLEASE SEND ME
A FULL AND DETAILED ACCOUNT OF EVERYTHING YOU
HAVE DONE STOP I WILL THEN BE IN A POSITION TO TELL
YOU WHAT TO DO NEXT STOP FOR THE PRESENT WOULD
ADVISE THAT YOU CONTINUE PRESENT TACTICS MAKING
EVERY EFFORT TO GET MISS HIGGINS ON OUR SIDE
ALEXANDER BOTTS

———

EARTHWORM TRACTOR COMPANY
EARTHWORM CITY, ILLINOIS
OFFICE OF THE SALES MANAGER

SEPTEMBER 1, 1932.

MR. ALEXANDER BOTTS,
METHODIST EPISCOPAL HOSPITAL,
VICKSBURG, MISSISSIPPI.

DEAR BOTTS: Your telegram is received, and I may say that I do not share your complete confidence this scheme of selling tractors to a hard-boiled dirt mover by working through his secretary. When you first proposed this idea, I was somewhat skeptical. But I realized that you had pretty thoroughly gone over the ground down here, and that you should, therefore, have a better grasp of the situation than a newcomer. So, with considerable reluctance, I decided to follow your advice. The results, up to this time, have not been reassuring.

However, I will follow your suggestion and give you a brief account of what I have done, and then I wish you to write me at once, explaining definitely all the facts which led you, in the first place, to believe that Mr. Slanker's secretary would have a deciding influence in this deal. And, after I have described the discouraging aspect of things here, I want you to let me know whether you still think it advisable for me to continue this line of action.

I arrived in Memphis day before yesterday—Tuesday—and spent practically all the afternoon at Mr. Slanker's office showing Miss Flora Higgins pictures and diagrams of our tractors and graders, and explaining their advantages.

Miss Higgins was cordial enough—in fact, she was almost too cordial—but she did not make a favorable impression on me. Just as you said, she is in full charge of the office here—not only that, she seems to be the entire office force—and she is young, and energetic, and good looking. But it is hard for me to believe that she would have much influence on someone who is thinking of buying a lot of expensive machinery. She just doesn't have the experience.

All the time I was showing the pictures and explaining the specifications, her attention kept wandering to other matters. About the only contribution she made to the subject under discussion was to say that she thought the tractor was "just too darling for words," and the grader was "as cute as a bug's ear." Most of her conversation had to do with her various boyfriends, the parties and dances she had attended in the past and was expecting to attend in the future, and her ambition to go to Hollywood and become a motion-picture actress. Finally, she invited me to go to a party with her that night. Not wishing to seem lacking in cordiality, I accepted—though not without considerable misgiving.

The party turned out to be even worse than I had feared. She brought along a group of her more or less half-witted boy and girl friends, and we put in a rather noisy but incredibly tedious evening at a distinctly low-grade speakeasy. Somehow, a man like myself, who is almost fifty years old, is married and has three grown children, seems to be out of place raising whoopee with the younger generation. However, I did the best I could. When the party broke up, everybody seemed to expect me to pay the bill, and so I did. It was $76.50, including tips.

Well, that would be all right, if it actually helped put over this deal. But I am not at all sure that it will. And my doubts are getting stronger all the time. Yesterday, and again today, I wanted Miss Higgins to come out with

Her attention kept wandering to other matters.

me to see a tractor demonstration which Fitz William has arranged to put on for her benefit. But I couldn't even get hold of her to tell her about it. Honestly, I am at a loss to know what she does with all her time—spends it with her worthless boyfriends, perhaps. She is hardly ever at the office. I called there twice yesterday and once this morning, and each time the place was closed up.

Finally, late this afternoon, I found her in. But all I got out of it was another fool invitation. She wants me to go on a party tomorrow night where, she says, I will meet members of some of the most aristocratic families in Memphis. She claims that she herself belongs to one of the important old families; she says they used to be very rich, but lost all their money during the Civil War. For this reason, she says, it is perfectly thrilling for her to meet someone like me who has so much money and is such a dear about spending it.

The more this woman talks, the less I like her. But she is so insistent about my attending this party that I have finally agreed. I only hope that you were right about her having the inside track with old Slanker; I would certainly hate to think I was sacrificing myself in vain. I can't imagine how this woman could have any legitimate influence over a man like Slanker, but maybe she had something on him, so that she can scare him into doing what she wants. I hope so, anyway.

Please answer this letter at once, and give me the complete and authentic dope on this business. Maybe you had better wire me; I'll go crazy if I keep up this foolishness much longer.

Yours,

GILBERT HENDERSON,

Sales Manager.

———

TELEGRAM

VICKSBURG MISS SEPT 2 1932
GILBERT HENDERSON
CARE EDWIN REGINALD FITZ WILLIAM
MEMPHIS TENN

DELIGHTED WITH YOUR LETTER STOP KEEP RIGHT ON
STOP YOU ARE GOING SWELL.

ALEXANDER BOTTS

———

CARE MR. E.R. FITZ WILLIAM
EARTHWORM TRACTOR DEALER

MEMPHIS, TENNESSEE.
SEPTEMBER 3, 1932.

MR. ALEXANDER BOTTS,
METHODIST EPISCOPAL HOSPITAL,
VICKSBURG, MISSISSIPPI.

DEAR BOTTS: Never, in all my experience in the tractor business, has anyone given me a line of advice so completely worthless as what you have been handing out the past few days. I don't like to be too hard on a man who is in the hospital, but a broken ankle does not excuse you for losing your brains. Every bit of information you have given me has turned out to be absolutely and completely wrong. And it is now up to you to do something—and do it quick—to get us out of the mess which you are responsible for. As a result of my believing your inaccurate statements and following your stupid advice, it now appears that we have got in so wrong with Mr. Slanker that this whole tractor deal may fall through. Incidentally, you have caused me a tremendous amount of annoyance and personal inconvenience, not to mention expense.

The party which Miss Higgins gave last night was much worse than the other one. We went to a speakeasy that was much more disgusting. The members of our party were much louder. And the bill was considerably higher—amounting in this case to $121.25, including tips. But even that was not the worst of it.

Immediately after I had paid this outrageous sum, the establishment was raided by the police, and all the guests, including myself, were taken to the station house, where we were forced to spend the few hours that remained of the night. In the morning, the judge, after giving us a most insulting talk, turned us all loose.

By slipping out in a hurry, I was able to shake off Miss Higgins. But later in the morning she called me up at Mr. Fitz William's office to request a loan of fifty dollars. She said that Mr. Jim Slanker had made a rush trip up from South Gumbo to find out what had happened to a number of important letters he was expecting. When he found that all his mail had been lying around the office for two weeks, he was pretty sore—which Miss Higgins considered most unreasonable. She admitted that it was her business to forward these letters, but she said he ought to realize that anyone is liable to make a mistake.

He also accused her of neglecting all her other duties, and she says she excused herself by telling him that she was looking after his interests by spending all her time with Mr. Henderson, of the Earthworm Tractor Company. His answer to this was a remark to the effect that "this guy Henderson had better not show up around here again; if he does, I'll knock his block off."

After making this statement, Mr. Slanker told her she was fired, paid her off, marched her out of the office, locked the door, and left for South Gumbo. Miss Higgins went on to tell me that, although she is perfectly furious at Mr. Slanker, she is, in some ways, rather glad to lose her job. She is tired of working, and this will give her a chance to take a little vacation trip up to St. Louis to visit one of her girlfriends who has recently married. It appears that the girlfriend's husband has money, so Miss Higgins thinks she can live on them very comfortably for weeks to come. But it will take a certain amount of cash to get up to St. Louis—hence the request that I lend her fifty dollars. In refusing this request, I made it pretty clear that we were through with each other forever. So I don't have to bother about Miss Higgins anymore—thank the Lord.

But there still remains the mess which she leaves behind here. And what a mess! This woman, whom you in your ignorance believed to be so important, and upon whom, following your advice, I wasted more than three days' time and more than two hundred dollars in cash, is now fired in a way that indicates she never did have any importance. And this is not all. In passing out of the picture, the dizzy dumbbell bleats around in such a foolish way that she contrives to get me in wrong, and make me look completely ridiculous in the eyes of this important prospect, Mr. Jim Slanker.

The whole thing leaves me in a quandary. I, naturally, hesitate to go down to South Gumbo at the present time. Any attempt which I might make to approach Mr. Slanker would be so handicapped by Miss Higgins's advance publicity about me that I should probably accomplish more harm than good. But something has to be done. And, as long as you are responsible for all this trouble, you will have to do it.

Broken leg, or no broken leg, you will have to straighten things out somehow. If you cannot go to South Gumbo yourself, you will have to get in touch with Mr. Slanker and fix it up in any way you can, so that I can go down there and meet him under favorable conditions. We must get his order.

I shall expect, by return mail, a full account of the steps you are taking to remedy your blunders.

Very truly yours,
GILBERT HENDERSON,
Sales Manager.

ALEXANDER BOTTS
SALES PROMOTION REPRESENTATIVE
EARTHWORM TRACTOR COMPANY

METHODIST EPISCOPAL HOSPITAL,
VICKSBURG, MISSISSIPPI.
MONDAY, SEPTEMBER 5, 1932.

MR. GILBERT HENDERSON,
CARE MR. EDWIN REGINALD FITZ WILLIAM,
MEMPHIS, TENNESSEE.

DEAR HENDERSON: Your letter, telling about the grand finale of your romance with little Miss Flora Higgins, is here. And it is a masterly effort. The other letter, telling about the beginning of the affair, was good too. But nowhere near as interesting and exciting as this last one.

I am sorry, however, that you have permitted yourself to get into such a worried state of mind about this tractor deal and everything. You take these things so seriously, and you seem to get all wrought up about them.

What you ought to do is try to act more like me. Here I have been flat on my back in a hospital for more than a week, and all the time I have maintained my usual bright and optimistic attitude toward life. My cheery wise cracks have been, and still are, a joy and an inspiration to all the nurses. On the other hand, you have been lapping up the night life of the great city, frisking about under the white lights, flitting from party to party, and in general acting the part of the little playboy of Memphis. Anybody would think you would be overjoyed at your good luck. But no, all you do is growl and complain.

And you have worked yourself into such a state of mind that you are accusing me of a lot of mistakes which I never made. As a matter of fact, I haven't made any mistakes at all. Every bit of the information I gave you has been correct. And all my advice has been perfectly sound.

For instance, I told you that I believed there might be someone behind the scenes in Mr. Slanker's organization who would have great influence over him in the purchase of equipment. This has turned out to be true. The power behind the throne is the mechanic who drove the tractor in the demonstration I started.

I never told you that Miss Flora Higgins had any drag with the old man. I merely advised you to stop off in Memphis and see her. And I stated the exact truth when, a little bit later, I wired you, "I have absolutely no doubts regarding Miss Higgins's influence on Slanker." Of course, I had no doubts. Mr. Slanker had told me himself that the office at Memphis was of no particular importance—being used only for the storing of records and the forwarding of mail—and that Miss Higgins was merely a girl whom he had hired cheap about a month before.

In advising you to call on this woman, I did not delude myself into thinking that you would accomplish anything constructive. I merely wanted to keep you pleasantly occupied in Memphis for a few days, and thus avoid the unfortunate consequences that would have resulted if you had persisted in going down to South Gumbo and inflicting your presence, white collar and all, on Mr. Jim Slanker and his operator. I will admit that my method of keeping you in Memphis was slightly disingenuous, but you must remember that you had already disregarded my previous direct warnings. I had to keep you out somehow. And it was lucky I did so.

I have just received a letter from Mr. Jim Slanker. He mentions his little trip up to Memphis, and then he says, "I just fired my stenographer. I find she had been neglecting her work to run around with some boob from the Earthworm Tractor Company by the name of Henderson. Apparently he is even more of a wet smack than that other guinea pig— Fitz William, or whatever his name was. It beats me how a company that makes such a swell tractor, and hires such a good man us you, could stand for any such washouts as these two."

So you see how it is, Henderson. If you could annoy and disgust old Jim to this extent without coming any farther than Memphis, it stands to reason that your presence at South Gumbo would have driven the old guy pretty near insane, and just naturally knocked all our chances on the head.

As it is, everything is working out swell. You had a taste of guy city life. Flora and her friends had a merry whirl with a guy that turned out to be a good spender. Old Jim and his operator had a chance to carry on the demonstration undisturbed, and my stay in the hospital was brightened up tremendously by those two charming letters of yours. In fact, the only trouble was that I laughed so hard while I was reading them that I almost fell out of bed and broke my leg all over again.

Mr. Slanker's letter, in addition to discussing his Memphis trip, states that he has completed the tractor demonstration, and that both he and his operator are completely satisfied. He has signed and mailed to me an order, which I had previously made out, for twenty-five sixty-horsepower tractors and ten elevating graders. He has added one condition to this order. He wants me to be there to take charge of the delivery of this machinery, and he specifies definitely that the Earthworm Tractor Company must agree, as part of the contract, to keep Messrs. Fitz William and Henderson out of his neighborhood from now on.

Very sincerely,

ALEXANDER BOTTS.

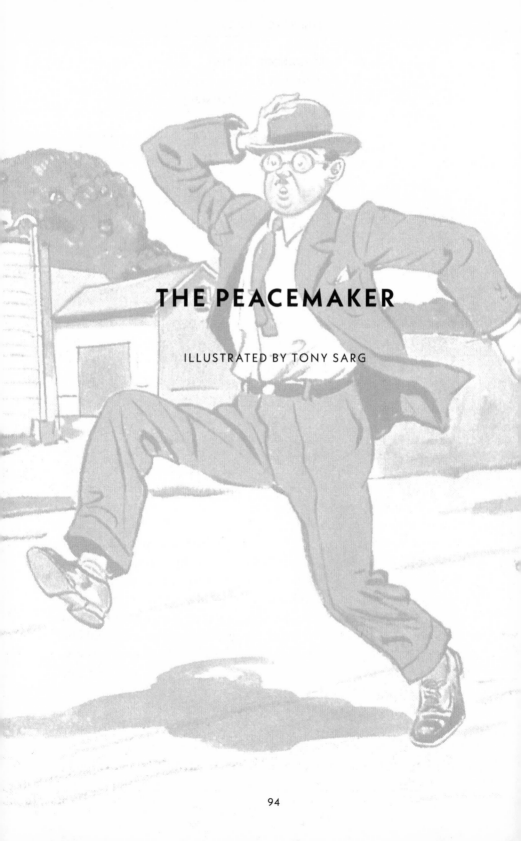

THE PEACEMAKER

ILLUSTRATED BY TONY SARG

ALEXANDER BOTTS
SALES PROMOTION REPRESENTATIVE
EARTHWORM TRACTOR COMPANY

HANGING GARDEN HOTEL,
BABYLON, MISSOURI.
THURSDAY EVENING, JULY 28, 1932.

MR. GILBERT HENDERSON,
SALES MANAGER,
EARTHWORM TRACTOR COMPANY,
EARTHWORM CITY, ILLINOIS.

DEAR HENDERSON: I arrived here this afternoon, called at once on our local dealer, Mr. Ben Garber, and ran into such an unfortunate situation that I have decided to stick around several days, if necessary, to straighten it out. I know that ordinarily you don't want me to spend that much time on a small dealer. But this is a special case—as you will realize when I give you the melancholy facts as set forth by Mr. Garber.

"Last spring," he told me, "I had a swell prospect by the name of Peabody, and another by the name of Snodgrass—both of them wheat farmers who actually had plenty of money. I was just on the point of selling each one of these men an Earthworm tractor and an Earthworm combined harvester. And then the Earthworm Tractor Company butted in and gummed the whole deal."

"How?" I asked.

"By sending out an idiotic pamphlet called 'Partnership Buying.' This pamphlet says that in many cases it is foolish for a farmer to buy a whole tractor all for himself—especially if he doesn't have enough work to keep it busy all the time. It says the thing to do is to have a group of farmers buy one machine in partnership. And it also says that when a tractor is sold to grain growers of established credit, the usual down payment should be waived and no money at all demanded until after the harvest. Honestly, Mr. Botts, I never saw such a bunch of hog wash. Have you read this pamphlet?"

"Yes," I said, "I have read it."

"Well," continued Mr. Garber bitterly, "Mr. Peabody and Mr. Snodgrass—my two prospects—read it too. It was sent out to the whole mailing list. And right away they fell for the idea. I told them it would never work. I told them they could never agree on which one of them would use the machine when, or who would pay for how much of what repairs. But they wouldn't listen."

"So you sold them a machine in partnership?"

"I had to, or else lose the sale entirely. They combined with a third wheat grower—a widow woman called Hopkins—and bought one outfit consisting of a tractor and harvester. Terms: Nothing down, and the rest after the harvest—if I can get it, which I probably can't."

"Why can't you?" I asked.

"Because they are all fighting like cats and dogs. Everything has gone wrong. Two of the owners—Snodgrass and Mrs. Hopkins—have already used the machine for harrowing and cultivating. It is now in my shop here being tuned up for the harvest. The wheat will be ready to cut next week. If it isn't cut on time, part of it may be lost—it's apt to fall down, or dry out and shatter. Each one of the three tractor owners will need six days to cut his wheat. So each one demands the machine for all next week."

"Doesn't the partnership agreement say when each one is to have it?"

"Oh, yes. They adopted the plan recommended in that lovely pamphlet. It is all to be decided by majority vote. But there is no majority. Peabody claims he should get the machine now because the other two have already used it. Mrs. Hopkins says her wheat is farthest along so she should be favored. And Snodgrass is a big, ill-natured brute who is just naturally stubborn and ornery. Each one of the three votes for himself, and says if he can't have the machine when he wants it, he won't pay one cent on it, and he'll sue the other two owners and me for damages."

"They're crazy," I said.

"Don't I know it?" said Mr. Garber.

"You could sue them," I said, "and make them pay. After all, they bought the tractor, didn't they?"

"Yes, but I can't afford a lawsuit. It would be too expensive—and bad for business. I can't afford to lose that money either. And I certainly can't afford to take the tractor back, now that it is secondhand. So there is only one solution. The Earthworm Tractor Company, by sending out that pamphlet, got me into this mess, and the company will have to get me out. It will have to take back this used machine and pay me the full list price for it."

"Holy Moses!" I said. "The company would never do that. Besides, it isn't necessary. All you have to do, Mr. Garber, is leave the whole thing to me. I will fix everything up."

"How?" asked Mr. Garber.

"By the use of diplomacy. Tomorrow morning I will go out and visit these three tractor owners. I will talk to them gently but firmly. By the use of remorseless logic I will show them that they are wrong. And by the use of tact and subtle suggestion, I will persuade them to get together in a reasonable way and work out a fair plan for the use of the tractor."

"What sort of a plan?" he asked.

"I will decide that as I go along. I will feel my way, and thus arrive at a perfect solution. They will then be completely satisfied. They will keep the machine and pay for it, and everything will be swell."

"It would be swell if you could do it. But you can't. I have already argued with these people till I am exhausted. And it does no good. They are all as stubborn as mules."

"Mr. Garber," I said, "for years my chief business with the Earthworm Tractor Company has been adjusting difficulties. I am an expert in the art of persuasion. I am a natural-born peacemaker. You just wait and see."

"All right," he said. "If you want to waste your time, it's OK with me."

He then told me how to find the three tractor owners. They all live on the same road to the north of here—Mr. Peabody nearest town, Mrs. Hopkins just beyond, and Mr. Snodgrass farthest out. I jotted down this information, and then wished Mr. Garber good afternoon and came over here to the hotel.

Tomorrow I will get busy on the diplomatic negotiations. Naturally, I can't expect to establish a permanent and lasting peace among three people who own one tractor. But I hope I can patch things up so they will get through the harvest and pay Mr. Garber. And, in the meantime, I would suggest that you suppress that pamphlet before it wrecks any more deals.

Incidentally, just who was the saphead that wrote this literary masterpiece? I wouldn't have supposed there was anyone in the Earthworm Tractor Company so dumb as to think that any group of people anywhere could successfully cooperate in the ownership of a tractor.

Yours as ever,
ALEXANDER BOTTS.

TELEGRAM
EARTHWORM CITY ILL JULY 29 1932
ALEXANDER BOTTS
HANGING GARDEN HOTEL
BABYLON MO

YOUR JOB IS PROMOTION OF NEW SALES STOP SUGGEST
THAT YOU DONT WASTE ANY MORE TIME ON MINOR
QUARRELS BETWEEN DEALER AND TRACTOR OWNERS
STOP THIS IS PRIMARILY THE DEALERS AFFAIR STOP
PARTNERSHIP BUYING IS A PERFECTLY SOUND IDEA
STOP I WROTE THAT PAMPHLET MYSELF STOP IF GARBER
USED A GOOD IDEA ON THE WRONG PEOPLE THAT IS HIS
RESPONSIBILITY NOT YOURS

GILBERT HENDERSON

———

ALEXANDER BOTTS
SALES PROMOTION REPRESENTATIVE
EARTHWORM TRACTOR COMPANY

HANGING GARDEN HOTEL,
BABYLON, MISSOURI.
FRIDAY, JULY 29, 1932.

MR. GILBERT HENDERSON,
SALES MANAGER,
EARTHWORM TRACTOR COMPANY,
EARTHWORM CITY, ILLINOIS.

DEAR HENDERSON: Your telegram is here. And it gave me something
of a jolt to learn that you wrote that pamphlet yourself. If I had known
this, possibly I would not have criticized it so freely. But you will understand that I feel no malice against you personally. As a matter of fact, I
was much pleased at the frank and honest way in which you admit that
you are the author. It is a great comfort to me to know that I am working for a man who takes full responsibility for his mistakes, and never
attempts to shift the blame onto anyone else.

I thoroughly agree with your idea that it is unwise to use up too much time on a small dealer like Mr. Garber. I will be on my way as soon as possible. However, I have got myself rather deeply involved in Mr. Garber's troubles, and it will be necessary for me to stay for another day, at least. I am sure that you will agree with me in this as soon as I have explained what I have done so far.

This morning—Friday—I rented a car at one of the local garages and drove out to visit the three partnership tractor owners. I called first on Mr. Snodgrass, who lives the farthest out. He turned out to be a very large, tough-looking baby—every bit as disagreeable as Mr. Garber had said. I found him just ready to start for town with a truckload of vegetables. When I announced that I wished to discuss the tractor situation, he told me, in a very vulgar manner, that he had no time to bother with a lousy city slicker. And, before I could explain that I was neither a city slicker nor infested with any sort of parasites, he drove off and left me.

At first I thought of chasing after him. But he seemed to be in such an unresponsive state of mind that I finally decided to let him go until some more auspicious time. Accordingly, I drove slowly down the road to the farm belonging to the Widow Hopkins. And here I met a very friendly reception.

Mrs. Hopkins is about thirty years of age and very good looking. Since the death of her husband, about four years ago, she has been making a very brave fight to run the farm.

As soon as she learned that I was from the Earthworm Tractor Company, she smiled at me very prettily, and said that she knew I must be a wonderful mechanic. When I admitted that I was not so bad, she insisted that I must look at an electric paint gun which had broken down just as one of her hired men was starting to use it to spray whitewash on some of the farm buildings. I immediately took the whole thing to pieces, cleaned and adjusted the working parts, and in less than an hour had it working perfectly. Of course, it was a very simple repair job, but Mrs. Hopkins thought it was wonderful.

"You don't know how you thrill me, Mr. Botts," she said. "I am always just overcome with admiration for a man who is strong and masterful, and knows how to do things."

This appreciation gave me such a warm glow of satisfaction that I decided to operate the gun myself. So I relieved the hired man and worked all the rest of the morning, whitewashing a large henhouse, two pigpens and a corncrib. Mrs. Hopkins said it was the best job of whitewashing she had ever seen.

So I relieved the hired man.

And the whole thing was a very clever move on my part. Mrs. Hopkins not only invited me to lunch—serving up the most succulent viands—but she was so pleased with me that she was ready to discuss the tractor situation on a basis of complete friendliness and mutual confidence. In consequence, I was able to uncover certain very important facts which Mr. Garber's clumsy efforts had failed to reveal.

In talking with Mrs. Hopkins, I first checked up the facts which had been given me by Mr. Garber, and I found that they were substantially correct. Mrs. Hopkins, in spite of her affable manners, was just about as stubborn as Garber had said. She was willing to cooperate, but only in case she could cut her wheat first. She realized that the other two owners were also demanding first whack at the machine, and she admitted that, if the problem was to be solved, somebody would have to give in.

When I suggested the possibility that we might work on Mr. Snodgrass, she said that she was completely disgusted with the man, and would prefer not to have any dealings with him at all. As I had just had a sample of Mr. Snodgrass's boorishness, I was inclined to agree with her. Accordingly, we passed on to a discussion of Mr. Howard Peabody. And at once I ran into some brand-new information.

"How do you and Mr. Peabody get along?" I asked.

"In general," she replied, "very well indeed. In fact, I am really very fond of Howard. We have known each other a long time. He is, on the whole, a gentleman. He is refined. He is genteel. And he is well-mannered. But just at the moment I am a little provoked at him."

"About this tractor business?" I asked.

"It is partly that. You see, he came over here yesterday, and he offered to back down and let me have the tractor to do my harvesting first."

"What? He offered to let you have the tractor first?"

"Yes."

"Splendid!" I said. "That means it is all settled."

"I'm afraid not," she said.

"Why not?" I asked. "You and Peabody are a majority. If you vote together, you control the situation."

"But I'm not going to vote with Mr. Peabody. I told him I couldn't agree to his plan."

"Why not?"

"Because I was provoked at him."

"I don't understand," I said. "You mean you got sore because he offered to let you have the tractor?"

"It wasn't that. If he had just offered me the tractor and stopped there, it would have been all right. It was what he said afterwards that I didn't like."

"And just what did he say afterwards?"

"I don't know whether I ought to tell you."

"You'd better. I'm trying to help you. So I have to know the facts."

"Well," she said, "if you must know, Mr. Peabody asked me to marry him."

"What?" I said. "He asked you to marry him?"

"I don't see why you have to act so surprised," she said. "Am I so repulsive that nobody would want me?"

"My dear lady," I said, "you misunderstand me. I should think everybody would want you. Why, if I was not already happily married and the father of twins—"

"Really, Mr. Botts—"

"What I mean," I said, "is that I was surprised to learn that Mr. Peabody had so much sense. But I still don't understand why you should be provoked at him, as you say. If he wants to marry you, why shouldn't he ask you? It doesn't hurt you any. You don't have to accept him unless you want to."

"Oh, I didn't object to his proposing to me. As a matter of fact, I was flattered. I was delighted."

"Well, what is the matter?"

"I didn't like what he said beforehand."

"And what did he say beforehand?" I asked.

"I just told you. He said he would let me use the tractor first. The two things went together. If I would marry him, he would let me have the tractor. If I wouldn't marry him, he said I could just whistle for the tractor—those were his very words. It was most insulting."

"I don't see why," I said.

"Of course it was insulting," she said. "He put the whole thing on such a low plane. What he was doing was asking me to sell myself to him for a paltry piece of machinery."

"It is not so paltry," I said. "An Earthworm tractor is a real masterpiece of engineering. But I can see how you feel about it."

"I feel very much hurt," she said. "If he had really cared anything about me, he would have offered me the tractor without any humiliating conditions."

"Mrs. Hopkins," I said, "I understand your attitude perfectly. And I am very glad you told me all the facts in the case. I feel that I am now in a position to solve all your difficulties. I will go right over and see Mr. Peabody, and I am sure that I can persuade him to let you have the tractor without any reservations or conditions."

"That would be just wonderful," she said. "Do you really think you can do it?"

"Trust me," I said. "I will go at once."

"But aren't you going to finish the whitewashing?" she asked.

"I thought I had."

"Oh, no. There is still the horse barn."

"Very well," I said. "I will whitewash the horse barn."

"That is really very dear of you," she said. "And I wish I could watch you do it. But I have arranged to drive over into the next county to visit some relatives. And I want to get started right away. I'll be back early tomorrow morning—in case you have any news for me. You won't mind if I go off and leave you?"

"That will be quite all right," I said.

I helped her get the car started, and, after she had driven away, I put in several hours squirting whitewash about the horse barn. I will have to admit that this irked me a little. It seemed to me that the widow was imposing upon my good nature a little too much. But I did not want to do anything that would endanger the *entente cordiale* which had sprung up between us. So I worked bravely along, and did a good job. Toward the end of the afternoon I finished up and drove over to call on Mr. Howard Peabody.

I found him just finishing the chores. He was alone; the hired man, who usually looked after the farm, and the hired man's wife, who took care of the house, were both away on a visit. Mr. Peabody was such a timid-looking little guy that I could not understand Mrs. Hopkins's fondness for him. Mrs. Hopkins, you will remember, had told me that she was much more partial to men like me, who are strong, masterful and can do things. It just goes to show that you never can tell about a woman.

I started in on Mr. Peabody by introducing myself as a representative of the Earthworm Tractor Company, and also an old family friend of Mrs. Hopkins. At first he seemed a bit suspicious. So I spent about fifteen minutes ladling out the old applesauce—telling him what a privilege it was to meet him, and what a really swell guy he was. This had the desired effect of convincing him that I was a person of rare judgment, and a man to be trusted. I then got down to the business in hand.

"Mr. Peabody," I said, "I have come to tell you that Mrs. Hopkins is very fond of you. When you asked her to marry you, she was highly flattered. She was delighted. She would have fallen on your neck—except for one thing."

"And what was that?"

"She misunderstood your attitude. She thought you were proposing an ignoble commercial deal by which you would swap the tractor for her hand in marriage."

"Well," said Mr. Peabody, in a rather whining tone of voice, "that was the idea, in a way."

"But not in the way she understood it," I said. "She actually believed you were asking her to sell herself, in a most sordid manner, for a piece of cold machinery. And I am sure, Mr. Peabody, you never even thought of such a thing."

"Well, not in those exact words, anyway."

"You see," I said, "it is all a misunderstanding. And it can be fixed up very easily."

"How?"

"You and I will go to town right away. We'll get the tractor and harvester from Mr. Garber. We'll bring them out here. And tomorrow morning when Mrs. Hopkins gets back from a little trip she is making, we'll take them over and deliver them to her with no conditions attached. This will give her a concrete proof of your unselfish devotion. Then, after she has used the machinery for a few days, you can ask her any favor you want and she will grant it."

Mr. Peabody was doubtful. "I would hate to lose the tractor and the lady both," he said.

"Listen," I said. "Does it stand to reason that any mere woman, if you give her half a chance, could hold out very long against a man like you?"

Mr. Peabody smiled. "I guess you are right," he said. "We'll go after that tractor right away."

"Now you're talking," I said.

We both got into my car. We started for town. And, as we drove along, I reflected that I had handled this thing pretty well. I had every reason to suppose that events were rapidly moving toward a successful conclusion. It soon appeared, however, that I was wrong.

As we approached the outskirts of town, we observed, rolling along toward us, an Earthworm tractor pulling an Earthworm combined harvester. As we passed the tractor, we observed that it was driven by Mr. Snodgrass. A few rods farther on I stopped the car.

"There is something funny going on here," I said. "What does old Snodgrass think he is doing with that tractor?"

"How should I know?" said Mr. Peabody.

I looked back over my shoulder. The tractor and the harvester were rolling slowly but steadily along the road.

"We might go back and ask him," I said, "but he is a rather hard guy to talk to."

"He is," agreed Mr. Peabody.

"If we want to get the real dope on this," I continued, "it might be better to talk to Mr. Garber."

I started up the car and speeded on into town. We found Mr. Garber in his office.

"Is that the partnership tractor that Mr. Snodgrass is driving out of town?" I asked.

"It is," said Mr. Garber.

"What's the idea?" I asked. "Who told him he could have it?"

"I did."

"Why?"

"Well, it's like this: He came in here a little while ago. He had been peddling vegetables around town all day. He made me a proposition. He said that if I would turn the outfit over to him so that he could do his harvesting first, he would pay me spot cash for his one-third share, and he would take his chances arguing it out with the other two owners. So I let him have it."

"You had no right to do that," I said. "Didn't you know that the use of the tractor was to be decided by majority vote? And didn't you know that I was out working on these people and fixing up an agreement with them?"

"Yes, but I knew you wouldn't have any luck. And a real cash payment in days like these looks awful good. I couldn't afford to let it get away."

"You poor, simple-minded oaf," I said.

"What?" he asked.

"You poor, simple-minded oaf," I repeated. "I had this thing all fixed up. And now you have to go and spoil it all. But it may not be too late yet . . . Come on, Mr. Peabody. Let's get after that guy and take the machine away from him."

"I'm afraid we can't," said Mr. Peabody. "And we want to be awful careful. Snodgrass is a big guy and he's pretty rough. If we bother him too much, he might beat us up or something."

"Come on!" I said.

I dragged him down the stairs and into the car, and started after old Snodgrass. We caught up to him just beyond Mr. Peabody's front gate. I drove right past him, parked at the side of the road, got out, and waved my hand for him to stop. He stopped. I walked up beside the tractor.

"My good man," I said, "I will have to ask you to get off that tractor. Mrs. Hopkins and Mr. Peabody—a majority of the owners of this machine—have come to an agreement regarding its use. Mrs. Hopkins is to have it first. Mr. Peabody second. And you third. Consequently, you will have to turn the machine over to me so that I can deliver it to Mrs. Hopkins. If you refuse, I warn you that we will have to use force."

During the course of this speech. Mr. Snodgrass occupied himself by glaring at me as ferociously as possible and interjecting various witless remarks that he had probably learned at the talking movies, such as, "Oh, yeah?" and, "Says who?" and, "You and what six other guys?" When I had finished, he merely opened the throttle and drove on. I followed along behind the tractor for a short distance, shouting out various remarks. But he paid no attention. Finally, I rejoined Mr. Peabody, who had remained seated in my car.

He was inclined to be sarcastic. "So you are the hero who was going to take the tractor away from the tough guy," he said. "It seems to me you are pretty weak. If I had been talking to him, I would have made him get right off that seat and walk home."

"Well, why didn't you?" I said. "You were right here all the time."

"I didn't want to interfere with you," he said, "after all the big talking you had done about how you were going to handle things. And now you haven't accomplished anything at all."

"Oh, yes, I have," I said. "There are more ways than one of getting a man off a tractor. What did you think I was doing when I was walking along behind the machine there?"

"It looked to me," he said, "as if you were handing out a lot of words that didn't mean anything to a man that wasn't listening."

"I was doing more than that," I said. "I was opening up the little drain cock in the bottom of the gasoline tank."

"What was the idea of that?"

"By this time," I said, "that tank is empty. In another couple of minutes the small amount of gasoline in the vacuum tank will be used up. The machine will stop. Mr. Snodgrass will discover that he has no fuel. And he will have to go off somewhere looking for more. Then we'll take charge of the machine and drive it away. But first we'll have to have some gasoline, and there may not be enough in this car. Have you got any at your place?"

"Yes," said Mr. Peabody. "But we can't do anything like this. Suppose Snodgrass should come back and catch us? He might get violent."

"All right," I said, "let him."

I swung the car around, drove back to Mr. Peabody's place, entered the front gate and stopped in front of the barn.

"I don't like this at all," said Mr. Peabody.

"Where's that gasoline?" I asked. "In the barn?"

"No," he said. "My insurance won't let me keep it in there. It's in the root cellar."

He led me around beside the barn to a sort of hatchway. We went down a flight of brick steps. Mr. Peabody unlocked a large padlock, swung open a heavy, insulated door, like the kind they use on ice boxes, and snapped on an electric light. We stepped into a fairly good-sized underground brick vault. The place was evidently intended for storing root crops and other vegetables, but at this season of the year it was largely empty. There were several baskets of peaches, several piles of old newspapers and magazines and a certain amount of miscellaneous junk. In one corner was a large drum of gasoline and a number of empty five-gallon oil cans. We filled up one of these cans.

"Now," I said, "we want to keep under cover until that guy gets out of the way. And we want to have some place where we can get a view of what's going on. How about the haymow?"

"All right," he said.

We carried the gasoline into the barn and set it in an empty stall. Then we climbed up a long, shaky ladder and floundered over the hay until we reached a small front window. From here we had a view of the road for quite a way in each direction. I was delighted to observe that the tractor and harvester were stalled about a quarter of a mile up the line.

"Look!" I said. "He seems to be walking around the machine. He is trying to find out what's the matter. Ah, ha! He's leaving. He's coming down the road this way. He'll walk right past here on his way to town."

"This is terrible!" said Mr. Peabody, in a weak little voice. "Maybe he's after us. Maybe he'll come right in here. I'm going to hide."

And then, believe it or not, the poor little shrimp began burrowing into the hay. In less than a minute he was completely covered up.

I remained at the window, cautiously peeking out. Mr. Snodgrass drew nearer. And, sure enough, when he reached the front gate he turned in.

"By the way," I said, speaking in a low tone of voice, "does this bum know where you keep your gasoline?"

A scared little whisper came up out of the hay. "Probably he does. He has been in here quite a few times."

"Did you lock up that cellar?" I asked.

"I'm afraid not. I was in too much of a hurry."

"Good Lord!" I said. "If he gets gasoline here, he'll start right back to the tractor, and we won't have time to make a getaway. I'm going down to lock that place up."

"You'd better not," whispered Mr. Peabody. "If he gets you, he might kill you."

"Shut up!" I said.

I crawled across the hay, went down the long, shaky ladder as fast as I could, hurried out of the barn and looked around.

There was no sign of Mr. Snodgrass. I rushed around the corner of the barn, leaped down the brick stairs, slammed shut the big door, pushed the hasp over the staple and snapped the big padlock in place.

Then, as I turned around and mounted the steps, there occurred a very peculiar phenomenon. I had a distinct feeling that I was hearing things. It almost seemed as if lusty shouts, muffled by the thick icebox door, were issuing from the depths of the root cellar. I began to wonder what had happened to Mr. Snodgrass, and, after a brief examination of the premises failed to reveal his presence, it suddenly occurred to me, as a remote possibility, that he might be inside the cellar. If this were the case, it would be my duty to let him out. I think I once heard somewhere that it is against the law to lock a man up without a warrant or something.

But, as I thought the matter over, I decided that I had no real reason for supposing that the man was in there at all. I had not seen him enter. I had not looked inside when I shut the door. And the faint sounds of shouting which I had thought I heard were probably a mere hallucination. It is well-known to scientists that even perfectly normal individuals, such as myself, may have slight temporary derangements of the functioning of the auditory nerve which give rise to such effects.

I decided that, on the whole, I had no responsibility in this matter whatsoever.

I had a perfect right to lock that door. I had done so with the consent of the owner. If Mr. Snodgrass had entered the cellar, it was for the purpose of stealing gasoline. If he got caught, that was his hard luck.

And it wouldn't do him any real harm, anyway. The place was large; he would have plenty of air. The temperature within was pleasantly cool as compared to the hot summer weather outside. He had a good electric light and plenty of old newspapers and magazines to read. If he got hungry, there were several bushels of peaches. What more could he ask?

I walked around in front of the barn. From here the muffled shouts were inaudible. I climbed up into the mow and finally succeeded in coaxing Mr. Peabody out of the hay.

"It's all right," I said. "The cellar is safely locked and Mr. Snodgrass has gone."

"Where? To town?"

"How should I know? He has gone somewhere to look for motor fuel. So it is up to us to get busy."

After much persuasion, I got Mr. Peabody down the ladder. We put the gasoline in the car and I drove up the road to the tractor. I primed the vacuum tank, closed the drain cock, poured the gasoline into the tank and then drove the tractor and the harvester off the road, across one of Mr. Peabody's fields and into a dense clump of bushes, where it was effectively hidden.

All this time Mr. Peabody had been very fidgety. He appeared to be in a panic for fear Mr. Snodgrass might come back and beat him up. At first I thought, of mentioning my vague suspicion that Mr. Snodgrass might possibly be locked up in the root cellar. But I decided not to. If he were really there, it would be, on the whole, better for our purposes if he remained there until after we had driven the tractor over to Mrs. Hopkins's in the morning. And I was afraid that if I mentioned the matter to Mr. Peabody he would insist on opening the door at once—on the theory that the big brute would be less dangerous at that time than after a whole night of captivity. It also seemed inadvisable to let Mr. Peabody spend the night at his farm. He might go into the root cellar after some of those peaches.

To guard against this possibility, I invited him to spend the night with me at the hotel in town.

My argument was that it would be unwise for him to stay at the farm alone. Mr. Snodgrass might show up at any moment, I said, and get violent. And I am happy to say that Mr. Peabody agreed with me.

So I brought him back to town with me. We had supper together. And since then, I have been writing this report in the lobby, while Mr. Peabody has been hiding up in my room.

So you see everything is going along fine. I have met with various difficulties, but I have overcome them all. And I expect to have everything finished up so that I can leave here on the noon train tomorrow—thus carrying out your desire that I do not spend too much time on these smaller dealers.

I will get this report on the night train, so that you will receive it in the morning.

> Very sincerely,
> ALEXANDER BOTTS.

TELEGRAM
EARTHWORM CITY ILL JULY 30 1932
ALEXANDER BOTTS
HANGING GARDEN HOTEL
BABYLON MO

YOUR REPORT IS HERE AND I DISAPPROVE HIGHLY OF YOUR WHOLE COURSE OF ACTION STOP IT IS BAD ENOUGH FOR YOU TO WASTE YOUR TIME PLAYING THE PART OF CUPID BUT WHEN YOU LOCK UP ONE OF OUR CUSTOMERS AND RUN OFF WITH HIS TRACTOR YOUR

CONDUCT IS HIGHLY DANGEROUS AND COMPLETELY UNWARRANTED AND INEXCUSABLE STOP YOU ARE HEREBY ORDERED TO SEE THAT SNODGRASS IS RELEASED AT ONCE AND YOU WILL THEN LEAVE BABYLON BY THE EARLIEST POSSIBLE TRAIN

GILBERT HENDERSON

———

ALEXANDER BOTTS
SALES PROMOTION REPRESENTATIVE
EARTHWORM TRACTOR COMPANY

HANGING GARDEN HOTEL,
BABYLON, MISSOURI.
SATURDAY NOON, JULY 30, 1932.

MR. GILBERT HENDERSON,
SALES MANAGER,
EARTHWORM TRACTOR COMPANY,
EARTHWORM CITY, ILLINOIS.

DEAR HENDERSON: What a day this has been! Yesterday was fairly exciting, but today has been positively frantic.

Early this morning I drove Mr. Peabody out to his farm. Everything was quiet and serene—except Mr. Peabody, who seemed a bit nervous. I avoided the neighborhood of the root cellar, and steered Mr. Peabody away from it. I stayed right with him while he fed the stock and did the other chores. Then I dragged him over to where we had left the tractor and harvester. I cranked up, we both climbed aboard and I drove out to the widow's farm.

When we arrived, I drew up beside the road, directly opposite the house, and right at the edge of a ten-acre wheat field. As I stopped, Mrs. Hopkins opened the door of the house and looked out. I cut off the motor, and told Mr. Peabody to run over and speak his little piece and ask the lady where she wanted us to put the machine.

So far everything had gone well, but just at this moment there arose a new and most annoying complication. As the roar of the motor died away,

I heard a distant noise as of someone yelling. Looking back, I saw a man running toward us along the road from the direction of Mr. Peabody's farm. The man was a couple of hundred yards away, but he was approaching fast. He was waving his arms and shouting angrily. I recognized at once who it was. It was my old friend Mr. Snodgrass.

It became necessary for Captain Botts to think fast. And I did. It occurred to me almost immediately that Mr. Snodgrass might be a bit irate, and that he might so far forget himself as to try to take the tractor away from us. I decided, therefore, that it would be just as well to fix the machine so he could not get very far with it. I leaped nimbly to the ground. I opened the drain cock in the fuel tank, just as I had done yesterday. And the remains of our five gallons of gasoline went splashing down over the rear end of the tractor. As I climbed back into the seat, Mr. Snodgrass drew up alongside of us. Just as I had feared, his state of mind was not exactly genial. If he had not been so out of breath from running, I believe he would have climbed up and tried to throw us both off the seat. As it was, he merely addressed a few remarks to Mr. Peabody.

"I've got you now, you rat," he said. "So you thought you could lock me up and steal my tractor, did you? Well, I fooled you. I cut my way through that door with my pocket knife. And now I'm going to have my innings. You're going to state's prison for most of the rest of your life for robbery and kidnaping. And I'll take that tractor right now. Get off that seat!"

"Really, Mr. Snodgrass," I said, curling my lip in a scornful smile, "I must ask you to keep your shirt in its accustomed place. We know nothing about your being locked up. And Mr. Peabody has a perfect right to this tractor. He will tell you so himself . . . Speak up, Mr. Peabody, and tell this guy where he gets off at."

But Mr. Peabody did not speak up. Unfortunately for him, he was on the side of the tractor nearest Mr. Snodgrass—so close to the huge brute that he was completely cowed. He just sat there, all drawn up like a scared rabbit and unable to emit even the most feeble squeak.

"Come on, Mr. Peabody, pull yourself together," I said. "Be nonchalant. Blow some smoke in the big bum's face, and show him you're not afraid of him." I handed over a cigarette and a box of matches. Mr. Peabody accepted them mechanically, put the cigarette in his mouth, lit it and threw the match back over his shoulder.

Well, it was just too bad. There was a sudden puff of flame behind us, and immediately the entire rear end of the tractor seemed to be on fire. Mr. Peabody leaped off one side of the machine, practically into the arms of Mr. Snodgrass, while I went bounding down the road and off into the wheat field. And this was not so good either. That wheat was dead ripe, and very dry. And I was no sooner in it than I found it was blazing and crackling furiously all around me. Two or three vigorous jumps got me out of the fire, but it spread so rapidly that I had to keep right on running to save myself from being burned up.

After I once got started, it was impossible for me to go back to the road. I had to continue straight ahead. And the fire, driven along by a moderate breeze, kept right behind me. I went up a small ridge, down the other side, and, finally, just as I was about to drop from exhaustion, I reached a back road which ran along the far side of the wheat field. Here the fire stopped, and there seemed little danger that it would spread in other directions. This particular ten-acre patch of wheat was gone, but the flames were held in on all sides by roads, and by woods and cornfields, which at this time were not dry enough to burn.

After resting myself for a few minutes, I started back. I attempted to circle around the still-smoldering wheat field. But I was so weary that my mind was a bit confused. While passing through a small patch of woods I must have lost my sense of direction. I staggered along through woods and cornfields for at least an hour, and finally reached a road which led me, not to the widow's farm, but directly into the little town of Babylon.

**Mr. Peabody leaped off one side of the machine,
while I went bounding down the road.**

In some ways this was rather fortunate, as it gave me a chance to go to the Hanging Garden Hotel, and get myself washed up.

When I entered the lobby, the desk clerk handed me your telegram. I opened it with great eagerness. In my somewhat discouraged state of mind, a rousing message of encouragement and confidence from my chief would have been a great help. You can imagine my disappointment when your communication turned out to be nothing but a good swift kick in the pants.

With greatly lowered vitality I wandered over to the office of our dealer, Mr. Garber. I hoped that he might have some news, and possibly a few words of good cheer. It turned out that he had some news, but absolutely nothing in the way of good cheer. Various reports about the fire had reached him from people who had come to town over the road which led past the widow's farm. And his nerves were very much on edge.

"So you are the guy," he said, "who called me a simple-minded oaf for letting Mr. Snodgrass take the tractor! And what do you do? You kidnap Mr. Snodgrass, who is one of my customers. You steal the tractor. And then you burn it up, along with a whole field of wheat belonging to Mrs. Hopkins, who is another of my customers."

"It was all an accident," I said.

"Accident or no accident," he said, "you and I are through with each other. You can just get out of this office, and out of town. And you can stay out. If you don't, I'll call in the sheriff and have you put in jail for arson and other misdemeanors and felonies."

"Very well," I said, "I'm on my way."

I withdrew in a dignified manner. And then I decided that I might as well go out and have a look at the remains of the tractor. By this time I was not as optimistic as I had once been as to the possibility of straightening out the affairs of the three tractor owners. But it seemed to me that it was still my duty to do anything that I could. I couldn't make things any worse, and there was a faint chance that I might be able to make them a little better.

I had a taxi man take me out to Mr. Peabody's farm. The owner was not there. I dismissed the taxi, climbed into the rented car—which I had left in Mr. Peabody's barn—and drove up the road. As I approached the Hopkins place, I decided to take no chances with Mr. Snodgrass. I would talk to him from the car, and I would keep the motor running so that I could make a quick getaway in case he got violent.

But when I arrived at the scene of the tragedy, there was nobody in sight but poor little Mr. Peabody. He was seated, the picture of dejection

I withdrew in a dignified manner.

and despair, on the charred and blackened seat cushion of the tractor. His eyes were fixed on the remains of what had once been a pretty nice field of wheat.

I drove up beside him.

"Good morning," I said. "Is Mr. Snodgrass around anywhere?"

"No," said Mr. Peabody. "He has gone to town."

I stopped the motor. And then Mr. Peabody started in to tell me what he thought of me.

"A swell friend you turned out to be," he said. "You were going to fix everything up." He smiled bitterly. "Well, I guess you've fixed it. But if old Snodgrass thinks he can put over anything as raw as this, he'll find out he's mistaken. I'll sue him. I'll get the money back."

"What money?" I asked.

"That check he made me give him. He has probably cashed it by now. But he can't get by with it. It was highway robbery. And Mrs. Hopkins came out here, and she just stood around and never lifted a finger to help me. She'll be sorry she acted this way. I wouldn't marry her now if she came to me on her bended knees."

"I don't understand," I said. "What happened?"

"I couldn't help myself," said Mr. Peabody. "Old Snodgrass claimed I started the fire."

"Well, you did, didn't you?" I said.

"You spread the gasoline all around. And you gave me the match. So it's really your fault. But Snodgrass blamed it all on me. He said I was the one who had ruined the tractor, and I would have to stand the loss. He made me buy out his share and Mrs. Hopkins's share. He took me by the neck and shook me all around. So I had to do what he told me. I wrote him a check for the full price of the tractor and the harvester."

"And what did he do with it?"

"He took it to town. He said he was going to cash it right away and pay Mr. Garber in full."

"What?" I said. "Mr. Garber is getting paid in full? Say, this is wonderful. This is marvelous."

"Yes. Just as marvelous as finding a dead cat in your soup. I suppose I ought to be giving three cheers and waving a flag. I pay the full price, and I get a burned-up wreck."

"Maybe it is not so bad," I said. "Let's look the old baby over."

I got out of the car and made a rapid inspection. The harvester was untouched. I looked over the tractor. The seat cushion and the paint at the rear end had been pretty well ruined. Aside from this, there was no

"He took me by the neck and shook me all around."

damage at all. It takes more than a small gasoline fire on the outside to do any real damage to a machine that is almost entirely steel and iron.

I drained several gallons of gasoline out of my car and put them in the tank. Then I cranked up the tractor motor. It ran fine.

"Mr. Peabody," I said, "there is nothing the matter with this machine. Fifty cents' worth of paint and a couple of dollars for a new seat cushion will make it just as good as new. You are a lucky man, Mr. Peabody."

"Well," he admitted, "I seem to be a lot better off than I thought I was."

"You are in fine shape," I said. "With a farm like yours you need a tractor and harvester all for yourself."

"I guess maybe that is right."

"Of course it is," I said. "So now you can just drive your machine on home. Thanks to me, you are sitting pretty."

"Thanks to you in a pig's eye," he said. "If I'm sitting pretty, it is due to pure bull luck. After all the monkeyshines you have been pulling around here, it is a wonder we both aren't in jail. I can't imagine why I was dumb enough to even let you come in my front gate. But you can be sure I never will again. Goodbye."

With these words he started up the machine, swung around and drove down the road toward his farm. As I stood watching him, I heard a cheery voice behind me. "Well, well! If it isn't Mr. Botts himself, the big paint gun expert!" I turned around. It was Mrs. Hopkins. "I came out to see what was going on," she said. "Is Mr. Peabody's tractor all fixed up now? Is he going to have a big repair bill?"

"The machine was hardly hurt at all," I said. "The total repairs won't amount to more than two or three dollars."

"That's fine," she said. "So now everybody is happy."

"I am glad you are not sore at me, Mrs. Hopkins," I said. "And I can assure you that I am sorry your wheat got burned up."

"I am not worrying about the wheat," she said. "It was only ten acres. And I have a lot more. Besides, I have something else to think about that is much more important."

"You don't say?"

"Yes, Mr. Botts. And it's so important that I just can't keep it to myself. I'm going to be married."

"Are you sure?" I asked.

"Of course I'm sure."

"Well, that's fine," I said. "But, from what he told me, I sort of got the idea that Mr. Peabody had changed his mind."

"Mr. Peabody! That little shrimp! He has nothing to do with the case. I'm going to marry Mr. Snodgrass."

"What? The big gorilla?"

"Really, Mr. Botts. You do say the oddest things."

"Well," I said, "you change your mind so quick I can't keep up with you. Yesterday you said you didn't want to have anything to do with Mr. Snodgrass. You said you were completely disgusted with him."

"Of course I was disgusted with him—because he wouldn't pay any attention to me. Ever since he bought that farm and moved in two years ago, I have been just crazy about him. He is so big, so strong, so virile. He is a regular superman—the perfect answer to a poor widow's prayer. I kept trying to strike up a friendship with him, but he would never respond at all. And that is why I got disgusted with him. I didn't know then that he really wanted to be friendly, but that he was just too shy."

"He didn't seem very shy to me," I said.

"Oh, he's bold enough with men. But with women, Mr. Botts, he is just like a great, big, awkward boy. You know, he has been secretly in love with me all the time, but he was too timid to say so. He wanted to take my part in all these discussions about the tractor, but he was too bashful. So he finally got hold of the machine, and he was going to bring it out to me. Isn't it just too cute for words?"

"Yes," I said, "a very whimsical situation indeed."

"Even then, he wouldn't have dared tell me how he felt about it. So it is probably just as well that you and Mr. Peabody went crazy and stole the tractor and everything. Because when he saw my wheat field burning up, he was so sorry for me and so angry at Mr. Peabody that he forgot all his bashfulness. Oh, it was wonderful, Mr. Botts. I was so proud of the masterful way in which he handled the situation. He shook Mr. Peabody as a terrier shakes a rat. He made him pay for the machine. Then he was going to make him pay for the wheat. He was just like a raging lion. But when I said that we did not want to be too hard on Mr. Peabody, he deferred to my wishes as gently as a lamb. It just shows how much power I have over him. And it certainly gave me a thrill. And then he came in the house, and his bashfulness held off long enough for us to get engaged, and then he took the check and went to town in my car, and we're going to be married next week and—look! Here he comes now."

I glanced down the road. A car was rapidly approaching.

"Well," I said, "I guess I'll be on my way. Goodbye and good luck to you."

As I climbed into my ear, Mr. Snodgrass drove up, leaped out of the car and walked over to Mrs. Hopkins.

"It is all fixed up," he said. "I paid Garber for the old tractor and harvester, and I've bought a brand-new tractor and harvester for us. It will be delivered tomorrow." At this point he caught sight of me. "So you are the man," he said, "that helped Peabody lock me up and steal the tractor and burn up Mrs. Hopkins's wheat! I've a good mind to knock your block off."

He advanced toward me in a menacing manner, but Mrs. Hopkins held him back.

"Don't hurt poor little Mr. Botts," she said. "He means well. And most of the time he's harmless—in spite of the fact that he seems to be crazy."

"All right," said Snodgrass. "But he had better get away from here before I change my mind."

"I'm leaving right now," I said. I turned the car around and started for town.

And, as I drove along, I couldn't help but congratulate myself on the complete success of my activities. Think of it—a bitter fight settled, a very pretty romance started and two complete tractor-harvester outfits sold where only one had been partially sold before. I tell you. Henderson, it just makes me feel good all over.

As ever,

ALEXANDER BOTTS.

ANIMAL BUSINESS

ILLUSTRATED BY TONY SARG

EARTHWORM TRACTOR COMPANY
EARTHWORM CITY, ILLINOIS
OFFICE OF THE SALES MANAGER

SATURDAY, MARCH 25, 1933.

MR. ALEXANDER BOTTS,
KANSAS CITY, MISSOURI.

DEAR BOTTS: When you get through at Kansas City, you will proceed at once to Prosperity, Missouri. Our dealer in that place, Mr. James Brown, has recently failed, and we have arranged to take over his business and run it as a factory branch. We will send out a permanent manager in a week or two. In the meantime, we want you to take charge, and we have written Mr. Brown to turn everything over to you.

Very sincerely,
GILBERT HENDERSON,
Sales Manager.

———

ALEXANDER BOTTS
SALES PROMOTION REPRESENTATIVE
EARTHWORM TRACTOR COMPANY

PROSPERITY, MISSOURI.
THURSDAY, MARCH 30, 1933.

MR. GILBERT HENDERSON,
SALES MANAGER,
EARTHWORM TRACTOR COMPANY,
EARTHWORM CITY, ILLINOIS.

DEAR HENDERSON: Your letter reached me in Kansas City on Sunday. I left there as soon as possible, and arrived here on Tuesday morning. It certainly was a happy thought on your part to give me this job. Already—in the course of only three days—I have converted Mr. Brown's moribund little business into a progressive enterprise that fairly seethes with activity.

Tuesday was a fairly quiet day. I spent most of the time with Mr. Brown, checking inventories and taking over the property, which consists of a brick building on the main street, with an office and showroom in front and a repair shop in the rear. There is one fifty-horsepower Earthworm tractor in stock; also a small supply of spare parts, and a fair amount of shop equipment and the usual office furnishings.

Mr. Brown is a pessimistic bozo, and he gave me a very gloomy account of local conditions. It appears that this town was named Prosperity back in the days of William McKinley and the full dinner pail. But right now it ought to have a name more suggestive of the wreck of the Hesperus, the sinking of the Titanic or the last days of Pompeii.

Nobody in the whole place, according to Mr. Brown, has any money at all anymore. The local bank has been closed for six months. Half the stores are closed. The rest are just barely able to keep going by bartering and trading with the farmers. During the past year Mr. Brown has not sold a single tractor, and his spare parts and repair business has amounted to only $29.52. Another tractor dealer—a man by the name of Clinton W. Montgomery, who handles the Emperor tractor, and who has a showroom a short distance down the street—has had even less success.

Mr. Brown assured me that I would not be able to do any business at all. At my request, he gave me the names of a half dozen people who had shown some interest in our tractor, but he said that none of them could be considered prospects, as they were totally devoid of funds.

We completed the work of going over the inventory and transferring the property late Tuesday, and Mr. Brown departed for Chicago, where he hopes he may be lucky enough to find a new job.

The next morning—which was yesterday, Wednesday—I began my first day as manager by inaugurating an entirely new policy, especially designed to overcome the adverse conditions which defeated Mr. Brown, and to make it possible for us to do business as briskly as in the good old days. In working out this policy, I first of all decided that I could not hope to sell tractors in this town for cash. The inhabitants have no cash. But they must have property of some kind. Therefore, it might be possible to sell tractors by barter. Many of the local merchants have already tried barter on a small scale. It has been successful. "Very well, then," I said to myself, "I will try out this plan on a large scale."

Accordingly, yesterday morning, I started out and visited the men who Mr. Brown said might be interested in tractors. They were all farmers—which means they were the type of people which I understand so thoroughly that I can mold them as wax in my hands, and make them believe almost anything. I told each one of them that he absolutely could not afford to be without an Earthworm tractor, and that I would supply him with a machine and accept in payment any sort of property, real or personal, tangible or intangible, which had a fair appraisal value equal to the list price of the tractor. I assured all these farmers that it was their patriotic duty to take advantage of this generous offer. I said that barter was the only means by which we could get business moving once more. And I exhorted them to put their shoulders to the wheel, so that we might lift the noble chariot of our economic life out of the slough of despond and bring it sailing triumphantly under full sail into the beautiful harbor of a bigger and better prosperity than ever before.

Everyone was deeply impressed by my words. But I did not want to hurry any of them, so I did not attempt to close any deals. I merely told them to think the matter over and, if they were interested, to see me later.

By Wednesday night I had visited all of the six men whom Mr. Brown had mentioned, and four or five others.

This morning—Thursday—after an early breakfast at the hotel, I went over to the office to wait for any possible developments. And it was not long before my efforts of yesterday began to bear fruit. A little before nine o'clock the door opened and in walked Mr. George N. Willow, one of the farmers whom I had interviewed the day before. He said he would like to trade in some of his farm animals for a tractor, and he wondered if I would care to look them over. I replied that I would be delighted. We both got into his car and drove out to his farm, about a mile from town. We spent several hours inspecting the livestock and discussing terms.

Mr. Willow is a very pleasant and thoroughly worthy fellow. Like many farmers, however, he is somewhat simple-minded, and inexperienced in business affairs. Consequently, in driving a bargain with him, I was able to obtain most favorable terms. He agreed to take the fifty-horsepower tractor which has been in stock here. And, in return, he gave me twelve splendid mules, which he values very conservatively at three hundred dollars apiece, and also two wonderfully fine cows, worth one hundred dollars apiece.

Of course, I rather hated to impose upon Mr. Willow's guileless innocence by taking away from him thirty-eight hundred dollars' worth of livestock for a tractor listed at only three thousand dollars, but I reassured myself with the thought that we deserve an extra profit for the very real service we are doing our customers by agreeing to transactions by barter instead of cash.

As soon as we had come to an agreement, Mr. Willow and a couple of his hired men helped me bring the animals in here to the office, where we installed them in the shop at the rear. Mr. Willow had very kindly included a week's supply of feed—hay and corn—so there will be no additional expense while the animals are held for resale. As soon as the livestock was settled in its new home, I turned over the tractor to Mr. Willow, and he drove it out to his farm.

I then sat down and wrote letters to various livestock dealers in Kansas City, St. Louis and Chicago, asking them to quote their best prices on high-grade mules and cows. I am now awaiting their replies.

Before I close this letter, I wish to state that I am much pleased at the way this new idea of mine is working out. During the entire past year, Mr. Brown did only $29.52 worth of business. In the course of less than three days, I have done more than one hundred times as much. I will keep you posted as to my future progress. Perhaps you had better ship me three or four more tractors to keep on hand. There is no telling what I may do in the next week or so.

Yours as ever,
ALEXANDER BOTTS.

———

TELEGRAM
EARTHWORM CITY ILL MARCH 31 1933
ALEXANDER BOTTS
PROSPERITY MO

IT IS NOT THE POLICY OF THIS COMPANY TO DO BUSINESS BY BARTER STOP YOU WILL SELL ALL LIVESTOCK AT ONCE AND CONFINE YOURSELF IN THE FUTURE TO TRANSACTIONS BASED ON CASH OR APPROVED CREDIT STOP WE DO NOT FEEL JUSTIFIED IN SHIPPING YOU ANYMORE TRACTORS UNTIL YOU TURN UP SOME REAL CASH CUSTOMERS

GILBERT HENDERSON

ALEXANDER BOTTS
SALES PROMOTION REPRESENTATIVE
EARTHWORM TRACTOR COMPANY

PROSPERITY, MISSOURI.
MONDAY, APRIL 3, 1933.

MR. GILBERT HENDERSON,
SALES MANAGER,
EARTHWORM TRACTOR COMPANY,
EARTHWORM CITY, ILLINOIS.

DEAR HENDERSON: I was much disappointed at the lack of warmth in your telegram of last Friday. And I have delayed answering it in the hope that I might be able to send you good news to the effect that I had sold all of the livestock at a handsome profit. It now begins to appear, however, that we cannot hope for any such simple solution.

I have received letters from all the livestock dealers to whom I wrote in Kansas City, St. Louis and Chicago. In describing market conditions, they say that mules are weak, soft and abnormally depressed. They even make the absurd claim that there is no such thing anymore as a three-hundred-dollar mule. From the descriptions I gave, they think they could pay about fifty dollars per mule, and possibly twenty-five dollars per cow. Naturally, I have not even answered the letters. I would not sacrifice these noble animals at any such prices. On the other hand, the alternative of keeping them here indefinitely has certain drawbacks. Fortunately, the expense is low. When I first made this deal, I resolved to save as much of the company's money as possible by caring for these animals myself. And I want you to know that I intend to go through with this plan, come what may. But I will have to admit that I am finding the work decidedly irksome.

As I think I explained to you in a former letter, I am keeping the live-stock in the shop at the rear of the tractor office. The place is of ample size—large enough to accommodate all the animals most comfortably. But, as there are no stalls, it is necessary for me to tie them up, here and there, to various work benches, drill presses and other more or less solid pieces of equipment. Unfortunately, the strangeness of the surroundings and the somewhat helter-skelter aspect of the accommodations seem to cause a certain morbid restlessness in these supposedly normal beasts. And, as their yearning for movement is not restrained by the side walls of the stalls to which they have been accustomed, they have a tendency to leap about, up and down, and particularly sidewise, in a way which makes me fear they may fall down, get tangled up in the tie ropes, or otherwise seriously injure either themselves or any of the shop equipment with which they may happen to collide.

In view of these facts, it has not seemed wise to leave them alone at any time; and I have, therefore, had the hotel move over a bed so that I can be on the job by night as well as by day. As there is not enough room in the office, and as the shop is already somewhat crowded, I have had the bed set up in the showroom right in front of the big plate-glass window. In this bed, each night, I have managed to snatch a certain amount of fitful and disturbed slumber. And from this bed, each morning, I have been compelled to arise at a most unholy hour—practically at the crack of dawn. On one occasion I overslept, and when I awoke I discovered that I was at the center of interest of a large crowd of yokels who had gathered on the sidewalk, and were looking in at me under the impression that I was some sort of window display.

My eating has been almost as difficult to arrange as my sleeping. At first I had the meals sent over. But recently, in return for a small tip, I have prevailed upon the hotel bell boy—a bright lad by the name of Bobbie—to come in and watch over my flock while I patronize the hotel dining room.

So far, my personal relations with the animals have been fairly good—with a few minor exceptions. One of the mules, who goes by the name of Phelps, is entirely too energetic, which makes him hard to handle, and there is one called Bill that is a mean little brute and is always trying to bite me or kick me. The one called Ben, however, is the worst. He has a very remarkable voice, and keeps bursting forth in song at all hours of the day and night.

But the thing that is wearing me down the most is the actual physical work. It is perfectly amazing how much time and energy I have to spend in merely dragging around the corn and hay to feed these brutes. Besides this, I have to water them. Then—as I want to keep them in good condition—I feel it my duty to brush them off every day. And, in addition to everything else, it takes a tremendous amount of activity with broom and shovel to keep the premises properly cleaned up. Sometimes I think I will have to devise a vacuum cleaner suitable for stables.

The work connected with the mules is bad enough. But when you consider that the cows have to be milked twice every day, you can well appreciate the sacrifices I am making for the good of the Earthworm Tractor Company. My fingers are so stiff I can hardly write this letter.

But I do not want you to think that I am in any way discouraged. As a matter of fact, the difficulties of this situation have merely stimulated me to greater efforts. Under the spur of necessity, my brain has been working even harder than my hands. And, as a result of the most intense mental activity. I have evolved a sensational but thoroughly practical plan which will not only solve all our present difficulties but will also provide a method by which we can successfully handle an indefinite number of similar ventures.

I am so impressed by this new idea of mine that I have already taken the first steps necessary to put it into actual operation. I have interviewed the receivers of a bankrupt canning factory here in town, and have ascertained that they would be willing to sell the entire plant, including equipment, with a total valuation of at least fifty thousand dollars, for only ten thousand. This plant is at present equipped for the canning of vegetables, but at very small expense it could be adapted for the canning of meat—thus providing a means by which we can utilize the animals which I have already taken in trade, and also making it possible for us to stimulate tractor sales all over the Middle West by going in for barter on a large and magnificent scale.

The possibilities are limitless. By taking in horses and mules at wholesale, and then subdividing them so that they can be sold at retail, we ought to make enormous profits.

To begin with, the best market for our product would probably be La Belle France, where large numbers of the inhabitants already have the good sense to recognize horse meat as the wholesome and nourishing food which, as a matter of fact, it is. Later on, we could develop a market in this country. Any aversion which may exist to the eating of equine products is entirely unscientific, and is based on mere emotional prejudice. It ought to be easy, therefore, to overcome this aversion by a high-powered advertising campaign. The best form of advertisement would probably be a full page with a testimonial by some society leader, a good-looking picture and some such title as, "The Daintiest Member of the Junior League—And She Eats Horse Meat."

But these minor details can be worked out later. At present the important thing is to buy the factory and get started. I will be confidently awaiting your ratification of my plan, and the receipt of a certified check for ten thousand dollars.

Yours as ever,
ALEXANDER BOTTS.

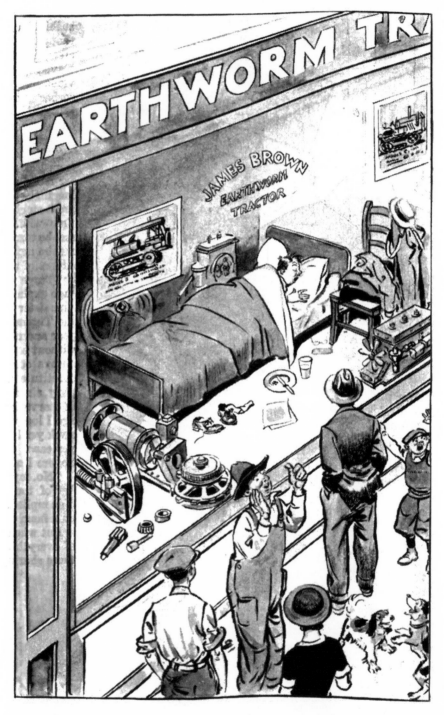

On one occasion I overslept, and when I awoke I discovered...

...that I was at the center of interest of a large crowd of yokels.

TELEGRAM
EARTHWORM CITY ILL APRIL 4 1933
ALEXANDER BOTTS
PROSPERITY MO

THIS COMPANY WILL NOT EVEN CONSIDER PURCHASE OF
CANNING FACTORY STOP WE ARE MOST EMPHATICALLY
OPPOSED TO SELLING TRACTORS BY BARTER STOP YOU
WILL SELL OFF ALL ANIMALS AS SOON AS POSSIBLE FOR
WHATEVER YOU CAN GET

GILBERT HENDERSON

———

ALEXANDER BOTTS
SALES PROMOTION REPRESENTATIVE
EARTHWORM TRACTOR COMPANY

PROSPERITY, MISSOURI.
WEDNESDAY NIGHT, APRIL 5, 1933.

MR. GILBERT HENDERSON,
SALES MANAGER,
EARTHWORM TRACTOR COMPANY,
EARTHWORM CITY, ILLINOIS.

DEAR HENDERSON: Your telegram arrived yesterday morning, and I
was naturally much depressed by your rejection of my splendid canning-
factory idea. I still think that in turning down this proposition you are letting slip one of the greatest opportunities in years. But the wording of your
telegram was so emphatic that I decided at once that further argument on
my part would be useless, and that I might as well accept the inevitable.

Accordingly, I started in at once to follow out your instructions in
regard to selling off the animals. I dipped into my carefully conserved
expense money sufficiently to have printed and distributed several hundred handbills advertising a sensational sale. The handbills were distributed yesterday afternoon and the sale took place today.

In preparation for the event, I subjected all twelve of the mules and
both of the cows to an unusually thorough brushing and currying. And

But the thing that is wearing me down the most is the actual physical work.

when I got through with them they looked as sleek and beautiful as anyone could wish. I moved them all into the front showroom, and placed the two handsomest mules, Adolph and Percival, right next to the big plate-glass window.

As a result of my publicity and my attractive window display, a great crowd of people attended. And, by evening, I had sold off the whole fourteen animals for more than their estimated value of thirty-eight hundred dollars.

This happy result was marred by only one small drawback. As none of my customers seemed to have any money at all, I was forced to accept as payment a somewhat miscellaneous assortment of property rather than the cash which I would have preferred. In spite of this, however, the sale was a pronounced success; my assets are now greater than ever, and they are in a form which will make them much easier to take care of while I am waiting to turn them into cash.

The most important purchase of the day was made by Mr. Clinton W. Montgomery. I think I mentioned, in a former letter, that this gentleman has an office a short distance down the street, where he sells—or attempts to sell—the Emperor tractor. In addition, he owns a large farm at the edge of town where he has been engaged for many years in the breeding of very high-grade purebred Percheron horses. He is a horse trader on a fairly large scale. And, as soon as he heard of my sale, he came right in and bought eight of my mules, giving in payment five of the most magnificent pure-black Percherons I have ever seen. These horses have pedigrees like British earls, and Mr. Montgomery admitted that they are worth at least five hundred dollars apiece—or a total of twenty-five hundred dollars. He was willing, however, to sacrifice them for eight three-hundred-dollar mules—worth only twenty-four hundred dollars—because he has a certain amount of rough work on his farm for which the mules are better adapted than the much heavier horses. Thus, his necessity is my gain. A further advantage, from my point of view, is that five horses will eat less than eight mules, so I will save money.

I further improved my position by selling another of the mules for a lot of hay, oats and corn, which will feed all of my present stock of animals for a long time. The last three mules I traded off to various people for a remarkable group of property, including one secondhand 1927 sedan which runs perfectly, one electric refrigerator, twelve sheep, one very good secondhand 1928 radio, two swans and two antique spinning wheels.

My shrewdest bargain, however, was the trading of the two cows for two hundred and fifty hens. These hens will lay eggs, and thus bring in just as good an income as was provided by the milk from the cows—and with less trouble, because the hens yield their product naturally. You don't have to squeeze it out of them.

To sum up, I am now in possession of the following property—the valuations being most conservative:

5	Percheron horses	$2,500
3	Truckloads hay, corn and oats	350
1	Automobile, 1927 model	800
1	Electrical refrigerator, almost new	170
12	Sheep	150
1	Secondhand radio, 1928 model	65
2	Swans	10
2	Antique spinning wheels	30
250	Hens	125
	Total	$4,200

You will note that the assets listed above represent a gain of four hundred dollars over the mules and cows previously held, and a gain of twelve hundred dollars over the original three-thousand-dollar tractor. Furthermore, I have every reason to believe that this new line of goods will be much more marketable, as well as easier to take care of, than the older. I have wired livestock dealers and secondhand furniture men in Kansas City, St. Louis and Chicago, offering my entire stock for sale. The market for this new stuff cannot be so bad as the market for mules and cows. And it may be so much better that when we finally turn this material into cash, we may make an even greater profit than what we have made so far.

Yours as ever,
ALEXANDER BOTTS.

———

TELEGRAM
EARTHWORM CITY ILL APRIL 6 1933
ALEXANDER BOTTS
PROSPERITY MO

WHEN I ORDERED YOU TO SELL OFF YOUR LIVESTOCK I MEANT A CASH SALE STOP YOU WILL GET RID OF THIS NEW COLLECTION OF ANIMALS AND ASSORTED JUNK AS SOON AS POSSIBLE AND YOU WILL SELL EVERYTHING FOR CASH EVEN IF YOU HAVE TO TAKE A CONSIDERABLY SMALLER AMOUNT THAN THE HIGHLY FICTITIOUS VALUATION YOU HAVE PLACED UPON THE PROPERTY STOP MR JOHN DODGE WHOM WE HAVE APPOINTED AS PERMANENT MANAGER AT PROSPERITY WILL ARRIVE THERE SATURDAY AFTERNOON AND WE WANT YOU TO HAVE EVERYTHING CLEANED UP SO THAT WHEN HE TAKES CHARGE HE WILL NOT BE HAMPERED BY A LOT OF ZOOLOGICAL ASSETS

GILBERT HENDERSON

ALEXANDER BOTTS
SALES PROMOTION REPRESENTATIVE
EARTHWORM TRACTOR COMPANY

PROSPERITY, MISSOURI.
FRIDAY EVENING, APRIL 7, 1933.

MR. GILBERT HENDERSON,
SALES MANAGER.
EARTHWORM TRACTOR COMPANY,
EARTHWORM CITY, ILLINOIS.

DEAR HENDERSON: Your wire arrived yesterday. In reference to your suggestion that all the assets which I have acquired by barter should be turned into cash, I hasten to assure you that I have planned, from the very beginning, to carry on this bartering in such a way that we would end up eventually with a lot of real money. In every transaction so far, I have increased the value of our assets. And if you would only let me take a reasonable amount of time to complete this plan, I could work the thing out in such a way that we would make a handsome profit. But you don't seem to understand all this. And apparently it is useless to argue with you.

Consequently, as you are the boss, I have no choice but to carry out your orders and do my best to liquidate this entire proposition before Mr. Dodge, the new manager, arrives tomorrow afternoon. Just how this can be managed in the short time remaining—less than twenty-four hours—I do not know. I have received replies from my various telegrams to Kansas City, St. Louis and Chicago, and they are not encouraging. Purebred Percheron horses seem to be a drug on the market. Quotations on sheep and hens are too low even to consider. And there seems to be no market at all for swans, secondhand electric refrigerators, 1928 radios and similar articles. As none of the people in town here have the cash to buy anything, I am left in something of a quandary.

But I am not discouraged. I will keep meditating on the situation, and before long I am sure to get some new and remarkable idea. So all the pessimistic and excessive doubts which you have expressed in your telegrams are totally unjustified.

Incidentally, I have been much disappointed in the tone of your comments on my enterprise here. I am doing absolutely the best I can, and I have been putting in an incredible amount of hard work in a spirit of unselfish devotion to the interests of the Earthworm Tractor Company. I don't think you realize what a terrific nervous strain this barter business has turned out to be.

Just as I was congratulating myself that my new assortment of animals would be less trouble than the over-energetic mules and irksome cows, it begins to appear that almost any animal is a nuisance.

The physical work of caring for twelve sheep, five horses, two swans and two hundred and fifty hens is not quite so great as for twelve mules and two cows, but it is, nevertheless, extremely wearying. And I don't like the atmosphere in which I have to work; the whole place smells just like a stable. Furthermore, there are many unpleasant surprises. For instance, this morning I stepped out for a moment to observe a dog fight which was taking place in the street, and when I got back, I found that one of the sheep had wandered into the front office and gnawed a big hole out of my best hat. And yesterday, one of the supposedly docile Percheron horses got scared at something, broke its halter chain, loped clumsily into the showroom and kicked out the big plate-glass window.

Fortunately, I have been able to conclude a deal with the local hardware store by which I am trading the electric refrigerator for a new plate-glass window, completely installed. But it is all most annoying.

One of the sheep had wandered into the front office and gnawed a big hole out of my best hat.

And this afternoon—to add to my distress and bewilderment—I discovered that some of the people with whom I have been doing business have not been entirely candid with me. As I was coming back from lunch, I happened to meet Mr. George N. Willow, the pleasant but dumb-looking farmer who had swapped me the twelve mules and two cows for the tractor. I asked him how the machine was running. You can imagine my surprise when he told me that he no longer owned it.

"You mean you sold it to somebody else?" I asked.

"I didn't exactly sell it," he replied. "I used it to pay off the mortgage on my farm. The mortgage amounted to twenty-two hundred dollars. It was long past due. Mr. Clifford Montgomery, the man who owned it, was pressing me for payment. I didn't have the money, so I gave him the tractor instead."

"You mean you gave him a three-thousand-dollar tractor to pay a twenty-two-hundred-dollar debt?"

"Yes."

"Holy cat," I said, "he certainly played you for a sucker."

"Oh, no," said Mr. Willow. "Mr. Montgomery did me a great service. He is a very kind man. Probably you have met him. He breeds horses and also sells the Emperor tractor."

"I know," I said. "I have met Mr. Montgomery. But I had not heard that he had acquired your tractor. How did you happen to think of offering it to him?"

"I didn't think of it" said Mr. Willow. "It was Mr. Montgomery who got the idea. You see, he came over last week Wednesday, just after you called on me the first time. He asked about the mortgage, and he was a little sore when I told him I couldn't pay. I offered to let him have the twelve mules and the two cows that I later offered to you, and also a dozen more cows that I had over in the other barn. But of course he wouldn't take them."

"Why 'of course'?" I asked.

"Because they weren't worth enough. I owed him twenty-two hundred dollars, and I couldn't have sold all those animals for much more than a thousand, the way prices are today. But then I happened to mention the fact that you had just been around offering to trade tractors for livestock. And that seemed to give him an idea."

"You don't know how you interest me," I said. "What happened next?"

"Well, sir," said Mr. Willow, "Mr. Montgomery seemed to be thinking *very* deeply for quite a while. Then he put in a long-distance call, and after that he made me a proposition. It seems that the man he had been tele-phoning to was a customer of his over in the next county, who had plenty of money, and who was thinking of getting a tractor, but had refused to buy an Emperor machine because he thought the Earthworm was better. This man had just agreed, in his telephone conversation with Mr. Mont-gomery, to buy a three-thousand-dollar Earthworm if Mr. Montgomery would sell it to him for twenty-eight hundred. So Mr. Montgomery said that if I could get the tractor out of you on a trade, he would take it off my hands in payment for the mortgage. This sounded fair enough to me. Mr. Montgomery would make a profit, and I would make a profit. So I got hold of you the next morning and put through the deal. And then I took the tractor over to Mr. Montgomery, and he was very much pleased. He was also kind of surprised."

"Oh, he was surprised, was he?" I asked.

"Yes," said Mr. Willow. "When I told him that I got the tractor for only twelve mules and two cows, he said he could hardly believe it. He said that if you were such a good man to trade with as all that, he thought he would have to go down and see if he could slip something over on you himself. He said he thought maybe he could get rid of some of those great big overgrown Percheron horses that used to sell for fancy prices, but which he hasn't been able to give away lately. From the way he talked, I imagine he will come down to see you very shortly."

"Thanks," I said. "He has been here already."

"And did you do any business with him?" asked Mr. Willow.

"Yes," I said." I swapped him eight mules for five of his horses."

"Well, well, well," said Mr. Willow. "You know, Mr. Botts," he contin-ued, "you are the kind of man that I really admire. You are so helpful. And I certainly appreciate everything you have done for me. Are you going to be around here a long time?"

"I'm afraid not," I said.

"I'm sorry to hear that," said Mr. Willow. "What this town needs to lift it out of the depression is more greathearted gentlemen like yourself. But I mustn't take up too much of your valuable time. Good afternoon, and good luck to you."

Mr. Willow walked on, and I entered my office, shooed a half dozen hens off the desk, and sat down to think things over. I was naturally pleased at Mr. Willow's opinion of me as the man who was needed to help pull the town out of the depression. On the other hand, however, I was a little upset by the discovery that, in transacting business with these small-town yokels, I had been going ahead without at all times knowing exactly and completely what I was doing.

While I was still in this disturbed state of mind, the door opened and in walked Mr. Clifford W. Montgomery. He had an entirely new proposition.

"How would you like," he said, "to trade back those live Percheron horses for the eight mules?"

"I'm not doing any more trading," I said. "But I would just as soon sell you the horses."

"How much?" he asked.

"Twenty-five hundred dollars for the lot," I said. "Your own price."

"What do you mean—my own price?"

"You told me yourself," I said, "that the horses were worth five hundred apiece."

"Sure," he said, "just like you told me your mules were worth three hundred dollars apiece. You could have got that much for them in 1929. But not today. Besides, I can't pay you cash; I haven't got it."

"My price," I repeated, "is twenty-five hundred in real money."

"How about eight three-hundred-dollar mules, plus one hundred dollars?"

"Not a chance."

"All right," he said. "Would you take one thousand dollars in cash?"

"I thought you said you didn't have any cash," I said.

"I haven't right now," said Mr. Montgomery, "but I expect to receive some tomorrow in payment for a certain piece of machinery which I am selling."

"Imagine that," I said. "You are selling a piece of machinery. Well, suppose you come back tomorrow and bring all the money you get for this piece of machinery, and perhaps we can talk business. Right now I will have to ask you to excuse me. I have to go out and look after my animals."

As I walked back into the shop, Mr. Montgomery followed me and attempted to continue the discussion. But I had made up my mind that I did not want to make any agreement with this old and crafty trader until I had had time to think the matter over at leisure. Consequently, I stirred up the hens and the sheep to such an extent that the ensuing bleating and cackling completely drowned out all attempts at conversation. Mr. Montgomery soon left in disgust, and I returned to the office, where I have been writing this report.

It is now four p.m. and I will spend the rest of the afternoon in meditation, analyzing the situation, evaluating its component factors and exploring all possible plans for future action. At the present time, of course, I do not care to prophesy exactly what may be the outcome of all this mental activity. I will let you know about it later.

Yours as ever,
ALEXANDER BOTTS.

ALEXANDER BOTTS
SALES PROMOTION REPRESENTATIVE
EARTHWORM TRACTOR COMPANY

PROSPERITY, MISSOURI.
SATURDAY, APRIL 8, 1933.

MR. GILBERT HENDERSON,
SALES MANAGER,
EARTHWORM TRACTOR COMPANY,
EARTHWORM CITY, ILLINOIS.

DEAR HENDERSON: Things have certainly been happening today. I have made a number of very clever moves. And in order that you may understand how very clever these moves were, I will go back and describe the mental groundwork upon which my whole plan of action has been based.

Yesterday afternoon, after I had mailed my letter to you, I sat down in the office and put the whole power of my intellect upon the various problems before me. And in the course of my meditation, I hit upon one small fact which probably would have appeared entirely unimportant, immaterial and irrelevant to most people, but which, to my analytical mind, seemed worth investigating.

"Why," I asked myself, "does Mr. Clinton W. Montgomery, after getting rid of five Percheron horses on Wednesday, suddenly decide on Friday that he wants them back? Why," I continued to ask myself, "should this shrewd and experienced horse trader act in such a paradoxical manner?"

After concentrating on this problem for about an hour, the answer suddenly leaped into my mind.

"Obviously," I said to myself, "some entirely new and highly important factor must have entered the situation between Wednesday and Friday. And it is now up to me to find out what this factor may be."

Having got this far, I gave my mind a rest for a short time while I went out into the back room, gathered up several dozen eggs, took them down to the general store and swapped them off for a package of cigarettes and a new hat. I then returned to the office and resumed my meditations in the midst of a fog of tobacco smoke, which acted not only as a pleasant stimulant to thought but also as an effective antidote to the persistent barnyard aroma which has permeated every nook and cranny of the premises. And before another hour had passed. I came to the conclusion that when a man like Mr. Clifford W. Montgomery changes his mind, it is most likely because of some economic reason. I decided that he must have received some sort of information which had caused him to think the horses were more valuable than he had supposed. Possibly, someone had offered to buy them at an unexpectedly high price. If this were so, I ought to know about it. I resolved to inquire around a bit.

By this time it was six o'clock in the evening. I, therefore, discontinued my meditations, fed the livestock, and telephoned Bobbie, the bell boy, to come and watch the establishment. Then I walked over to the hotel.

By examining the register, I discovered that one new guest had arrived the day before—a Mr. Johnson, of St. Louis. The clerk was kind enough to point him out to me. I promptly introduced myself and proceeded to get acquainted. At first, Mr. Johnson was inclined to be reserved, but after I had plied him with cigars and told him a number of my best Swedish jokes, he began to warm up, and consented to have supper with me. In the course of the meal, I drew him out more and more, and finally discovered, much to my satisfaction, that I had been following a real live hunch.

Mr. Johnson represents a prominent brewer in St. Louis. This brewer, in the old days, used to take the greatest pride in the splendid Percheron horses which pulled his wagons, and which became so famous that they inspired some prominent composer to write a song about them called The Brewer's Big Horses.

"And now that beer has come back," explained Mr. Johnson, "this brewer is in the market for horses. He won't hear of motor trucks. He wants to be old-fashioned—partly because he is a sentimental old bird, and partly because he thinks it would be good publicity for his business. So I am up here negotiating with a man called Montgomery. He's an old horse breeder, and he used to sell us horses back in the good old prewar era."

"And how are you getting along with Mr. Montgomery?" I asked.

"All right, I guess," he replied. "My requirements are pretty rigid. The boss is fussy. He wants twenty-four purebred animals of the highest type—all black, and all just alike. Mr. Montgomery has nineteen that fill the bill perfectly, and he says he thinks he knows where he can pick up the other five. He will let me know tomorrow. If he can supply the whole twenty-four, he gets sixteen thousand dollars in certified checks, which I brought along. That's a lot of money. It's way above the market. But expense is no object to this brewer if he can get what he wants."

"Suppose," I asked, "Mr. Montgomery can't get these extra five horses?"

"Then the whole deal is off," he said.

"It is a very interesting situation," I said, "and I can't tell you how much I have enjoyed this little chat. It is very seldom, Mr. Johnson, that I meet a man who talks in such an interesting way."

Well, the rest was easy. When Mr. Clifford W. Montgomery showed up at the office this morning, I was all ready for him. But before I got started working on him, he spilled one more bit of news.

"I am now in a position," he said, "to make you a very fine offer for those horses. I have purchased the Earthworm tractor which you sold to Mr. Willow. And I am ready to give you this three-thousand-dollar machine instead of the twenty-five hundred dollars in cash which you mentioned yesterday. You see," he explained, "I got the tractor for a customer of mine over in the next county. He had agreed to buy it. But when I went over last night to close the deal, he told me he had changed his mind and bought a cheap secondhand machine. As the agreement was oral, I can't hold him to it. So I'm stuck. It is most annoying."

"Yes," I agreed, "and that is not the worst of it. Before I get through with you I am afraid you are going to be annoyed a whole lot more." I then repeated to him the conversation I had had with Mr. Johnson, the horse buyer from St. Louis. "And now," I continued, "you will under-stand that I cannot afford to renew the offer which I made—and which you so unwisely rejected—yesterday. However, if you really want the horses, you can have them for six thousand dollars in cash."

At this, Mr. Montgomery began raising such a holler that he disturbed all the animals in the back room, and they began such a bleating and cackling that it was impossible for some time to continue the negotiations. Even the two swans joined in with their peculiar characteristic cry.

Finally, however, things got quieted down. We talked the situation over, and out of pure generosity and good will, I relented slightly in my demands, and agreed to take the tractor and three thousand dollars in cash. The deal was completed an hour later with the cooperation of Mr. Johnson, who received the twenty-four horses and in return paid thirteen thousand dollars to Mr. Montgomery, and three thousand dollars to me. Mr. Montgomery then gave me the tractor.

This happy result made me feel so magnanimous that I at once decided to make a very substantial gift to the local Red Cross for the benefit of the unemployed. I turned over the twelve sheep, the two hundred and fifty hens, and two swans and the two antique spinning wheels. It was my idea that the unemployed could shear the sheep, convert the wool into yarn on the two spinning wheels and then knit themselves a lot of sweaters, socks and other garments. The man in charge was a bit doubtful about this, but he was glad to receive the property. I think he has a false idea that he can sell it somewhere.

After completing this transaction, I made a deal with a local vegetable grower by which he gave the premises a thorough cleaning, and received in return for it two large wagonloads of manure.

And late this afternoon, when Mr. Dodge, the new manager, arrived, I was able to turn over the business to him with everything in fine shape. As we sat in the office, and as I explained to him that he had in the treasury the three thousand dollars which I had received for the tractor I had sold, that he also had the tractor back again, all ready to sell a second time, and that he had a nice radio and a secondhand car—as I explained all this, and emphasized the fact that these benefits were due entirely to my shrewdness and efficiency, we were interrupted by a sudden disturbance outside.

Going to the door, we observed that the entire street seemed to be filled with large black Berkshire hogs. There must have been fifty or sixty of them, and they were driven along by an active and breathless farmer, assisted by four helpers. As we gazed upon this somewhat astonishing scene, the farmer came over and addressed us.

"I understand," he said, "that you people do business by barter. So I have brought in these hogs and I want to swap them off for a tractor."

As Mr. Dodge, the new manager, seemed too bewildered to make a reply, I took charge of the situation in my usual competent manner.

"My good man," I said, addressing the farmer, "we have decided to do no more business with farm animals. We are machinery men, pure and simple. But there is, in this town, a man who sells tractors, and who is also a very active livestock trader. He is a competitor of ours, but I am big-hearted, and I would be delighted to see him make a deal with you. I would suggest, therefore, that you conduct these animals to the place of business of Mr. Clifford W. Montgomery. Goodbye."

Mr. Dodge and I stepped back into the office, and the hog parade continued on down the street.

As ever,
ALEXANDER BOTTS.

COOPERATION

ILLUSTRATED BY TONY SARG

ALEXANDER BOTTS
SALES PROMOTION REPRESENTATIVE
EARTHWORM TRACTOR COMPANY

QUITTYQUINK HOTEL,
QUITTYQUINK COVE, MAINE.
MONDAY EVENING, JUNE 26, 1933.

MR. GILBERT HENDERSON,
SALES MANAGER,
EARTHWORM TRACTOR COMPANY,
EARTHWORM CITY, ILLINOIS.

DEAR HENDERSON: I arrived in this little seaport town this morning, and immediately ran into a situation which fills me with joy and hope. It looks to me, Henderson, as if the depression is definitely licked. And I want you to know that I am doing everything in my power to see that it stays licked. I believe that the best way to encourage and strengthen this new flood of prosperity is for all of us business men to work together, forgetting our old-time suspicion and distrust of each other, and putting our shoulders to the wheel like one great, happy family. I have been much disappointed to note that you boys at the home office have been very slow in getting together with the other tractor manufacturers and drawing up a code by which you can do business in a reasonable and public-spirited way. I trust that you will be inspired to move a little faster when I tell you of the wonderful venture in cooperation which I am initiating in this locality.

As soon as I arrived in town this morning, I followed my usual procedure of calling on the local Earthworm tractor dealer to find out how I could best assist him in pepping up his sales. Unfortunately, the dealer was sick in bed, so I decided to take entire charge of his business for a few days.

I started in by inspecting the office and showrooms, and was pleased to find that the dealer has in stock one of our latest seventy-five horsepower diesel tractors equipped with our new power-driven logging winch. My next step was to buzz about town and pick up information. I discovered that the lumber business in this region is definitely improving; before long, it may be possible for us to sell a lot of tractors for logging operations. I also discovered that we may have a certain amount of competition from one of the natives here—a very energetic gentleman by the name of Ezra Litchfield, who has the local agency for the Emperor tractor.

And then, toward the end of the morning, I ran into some really exciting news. Just by accident, I happened to meet a man by the name of Russell Jones, who is staying here at the hotel, and who told me a most interesting story.

Mr. Jones is an inventor, a designer of airships and the chief engineer of the Detroit All-Metal Dirigible Corporation. He is up here working on a deal by which his company expects to sell an airship.

The prospective purchaser is Mr. T. Henry Mott, a big millionaire from New York, who has an elaborate estate covering the whole of Quitty-quink Island, which is four miles long, two miles wide and lies just off the coast here. It seems that Mr. Mott, back in 1928, bought a small dirigible, or blimp, which he used for commuting back and forth from New York. In 1930, he lost a lot of money, and had to sell the blimp for reasons of economy. But he still has an adequate landing field on the island, and a large hangar.

During the last few months, his business has been improving so fast that he has decided to buy a new airship, and Mr. Jones's company has sent one on from Detroit to New York to give him a demonstration. The demonstration will be a trip from New York up here—leaving New York next Thursday morning, and arriving here sometime Thursday afternoon. Mr. Jones came on ahead to get enough men for a ground crew and make all arrangements so that the ship may be properly landed and put in the hangar. He is all in a dither about it, because this is the first chance they have had to sell an airship for years, and Mr. Mott has already told them that if he decides to buy it he will pay cash—which will mean exactly two hundred thousand dollars in their treasury. You can imagine my excitement when I heard this.

"Good Lord!" I said. "If people are actually buying two-hundred-thousand-dollar airships for cash, it must be that we are entering upon real boom times."

"He hasn't bought it yet," said Mr. Jones.

"If he is even considering such a thing," I replied, "it shows that times have changed. I should think, though," I continued, "that he would be better off with an airplane—which would probably be a lot cheaper, and also a lot faster."

At this remark, Mr. Jones looked very scornful indeed. Apparently, he is a nut on the subject of dirigibles. "Airplanes are all right for cheap sports," he said. "But a man of Mr. Mott's wealth would naturally prefer to spend a little more and get something really worthwhile. If he buys our ship, he will have a luxurious yacht with a large cabin, four attractive staterooms with wide and luxurious beds, ample kitchen and dining accommodations, with all accessories. The ship has a maximum speed of eighty miles an hour. So he will be able to come up from New York in less than five hours. This is fast enough. He can spend the evening in New York, going to bed any time he feels like it, have a refreshing night's sleep and awake in the morning up here. Or he can make the trip in a single morning or a single afternoon—cruising along in a pleasant and civilized manner, admiring the scenery, conversing with friends, reading books, enjoying the radio, dictating letters or occupying his time in any way he wishes."

"Couldn't he do that in a plane?" I asked.

"Certainly not," said Mr. Jones. "In a plane he would be bumped and bounced along over the air pockets, deafened by the roar of the motors, and half shaken to pieces by the vibration. His limbs would be numb and cramped from sitting in a small and highly uncomfortable seat, and the nauseating motion would have his stomach in a state of open rebellion."

"Well, well, Mr. Jones," I said. "You seem to have a sales talk that is as good as anything we tractor salesmen can hand out. I have ridden in airplanes quite a bit, and I always thought I was having a good time. But now that you have given me the real dope on the subject, I'll never be satisfied with anything less than a real airship."

"Any airship—even an ordinary one—is better than an airplane," said Mr. Jones. "And our all-metal ships are the best of all. They are, in fact, the last word in aeronautic construction. Instead of covering the aluminum framework with a fabric envelope, which is subject to leaks and rips, and which rapidly rots away on exposure to the weather, we use an envelope made of a new aluminum alloy which is as light as a feather, as tough as steel and can be rolled out into sheets of a paper-like thinness. Constructed of this splendid material, our ships are lighter and also stronger than the older fabric affairs; they are leakproof, fireproof, weatherproof, and practically indestructible. The cost of upkeep is practically nothing."

"You don't know how you interest me," I said. "Are there any of these all-metal airships in existence?"

"So far," said Mr. Jones, "there are only two in the entire world—the one which we are now trying to sell to Mr. Mott, and another somewhat smaller one which was bought some time ago by the United States Navy. But in ten years there will probably be as many of them in the air as there are now ships in the ocean."

"And don't they have any disadvantages at all, aside from the expense?" I asked. "Aren't they rather clumsy things to land?"

"Any large ship is clumsy in port," said Mr. Jones. "An ocean liner requires tugboats to bring it into the dock. And an airship needs a ground crew to haul it into the hangar. This is, of course, a certain disadvantage—especially up here, where I am having trouble in hiring enough people. To be perfectly safe, in case next Thursday is a windy day, I ought to have at least a hundred men. Quittyquink Cove is such a small town that I'm afraid I'll have to bring some in from outside, which will mean a lot of expense at a time when our company is trying to economize. Also, I may have some trouble in getting a hastily gathered gang of men to work together."

At this point, it suddenly came over me that Old Man Opportunity himself had arrived at the door and was knocking loudly. I did not keep him waiting. "I should think," I said, "that instead of a lot of puny human beings, you could use a large tractor to haul this airship into its dock."

"Of course, I could," said Mr. Jones. "But I would need a pretty big one, and I couldn't afford to buy it just for one landing. Besides, the ship is arriving on Thursday; it is too late now to send off and get a machine."

"Listen, Mr. Jones," I said, "I represent the Earthworm Tractor Company. I have a splendid, brand-new, seventy-five horsepower diesel Earthworm tractor right here in town. It is equipped with our latest three-speed reversible power winch. This is the finest tractor ever built in this or any other country. If you say so, I will take it over to the landing field and operate it as a tugboat to pull your big ship into its dock. I won't charge you a cent, and I will guarantee satisfaction. This tractor weighs more than a hundred men, and can pull harder than five hundred. By running a line through the winch, we can pull in or slack off at will. Our control will be perfect. What do you think of the idea?"

"It sounds fine," said Mr. Jones, "and it's most generous of you to do it for nothing."

"It is not generosity," I said. "It is cooperation. I assist you in making a good landing, and this will help you in selling your airship. You give me a chance to show Mr. Mott what my tractor can do, and this will help me sell him the tractor as a part of the necessary equipment for his landing field. By working together and helping each other, we each gain greater advantages than if we worked alone. Isn't it beautiful?"

"It sounds all right to me," said Mr. Jones. "We really ought to have two tractors, though—one for the bow and one for the stern."

"I am sorry," I said, "but I have only one machine available. I'm afraid you'll have to get along with that."

"Very well," he replied. "I accept your proposition, and I thank you."

We shook hands. Then I asked him to keep the matter a secret, and took my departure.

I did not tell Mr. Jones why I desired secrecy. The idea, of course, is that Mr. Ezra Litchfield, the local Emperor tractor dealer, has one large machine in stock, and I don't want him horning in on this party—especially as Mr. Jones had indicated that he would welcome the help of a second tractor.

In closing, I wish to emphasize the importance of this new cooperative movement which I am starting. By working hand in hand with the airship industry, I am opening up an entirely new market for Earthworm tractors. If Mr. Jones ever succeeds in his plan of putting as many airships into the air as there are now steamships in the ocean, we will be able to sell a truly fantastic number of tractors to tow them in and out of their houses when they come to earth.

I will keep you informed as to my progress.

Most sincerely,
ALEXANDER BOTTS.

———

ALEXANDER BOTTS
SALES PROMOTION REPRESENTATIVE
EARTHWORM TRACTOR COMPANY

QUITTYQUINK HOTEL,
QUITTYQUINK COVE, MAINE.
TUESDAY EVENING, JUNE 27, 1933.

MR. GILBERT HENDERSON,
SALES MANAGER,
EARTHWORM TRACTOR COMPANY,
EARTHWORM CITY, ILLINOIS.

DEAR HENDERSON: You will be delighted to learn that I am extending the scope of my cooperative efforts over a field much greater than that described in my letter of yesterday. I am now cooperating, not only with the airship industry, but also—believe it or not—with

Mr. Ezra Litchfield, the local selling agent of the Emperor Tractor Company. The idea of one tractor salesman actually cooperating with a rival tractor salesman is so novel, so unheard of, and so incredible, that I will have to explain the circumstances in some detail in order that you may understand why I embarked upon this remarkable enterprise, and so that you may appreciate the advantages which it will bring us. The whole thing came about because of an unforeseen difficulty which I ran into this morning.

Right after breakfast I began making inquiries as to the best way to take my tractor over to Mr. Mott's island. And you can imagine my shocked surprise when I discovered that there did not seem to be any practicable way to do this. There is no boat in this small fishing port capable of transporting a twelve-ton, seventy-five horsepower Earthworm tractor. And even if we brought in such a boat from somewhere else, there is no adequate pier or dock from which we could load the tractor onto the boat. The local seafaring men seem to think that I would have to drive the tractor half-way around Penobscot Bay to Belfast, and ship it from there. Besides being expensive, this might take three or four days—which would never do at all, as the airship is expected day after tomorrow.

In the hope of finding some way out of this dilemma, I went over and inspected the channel which separates Quittyquink Island from the mainland. At its narrowest point, it is almost a quarter of a mile wide. The bottom is nice, smooth gravel. But the water, even at low tide, is twenty feet deep, so it would be impossible to drive the tractor across.

As all of the more obvious methods for getting the tractor to the island were useless, I got the old bean working, and finally cooked up a plan of action which is somewhat unusual, but which seems to be the only way in which I can successfully overcome the difficulties of this somewhat peculiar situation. To carry out my plan, I would need two tractors, and this presented a momentary difficulty, owing to the fact that I had only one. The only other tractor in town was the Emperor machine belonging to Mr. Ezra Litchfield. I had hoped to keep this gentleman out of the affair entirely. But I absolutely had to have his tractor, so I went around to his office and generously offered to let him cooperate with me.

Mr. Litchfield, I am happy to say, turns out to be a very high-grade individual. I have had a great deal of experience in sizing up men, and I could tell from the virile way in which he shook hands, and from the frank and steady way he looked me in the eye, that he is a man of rugged honesty and great strength of character. Besides being the local Emperor tractor dealer, he holds the office of county sheriff, which shows that he is well regarded in the community. And he is a worthy son of illustrious forbears, being descended from a long line of New England sea captains—sturdy souls who, in days gone by, carried the Stars and Stripes to every seaport in the world and made the name "American" synonymous with courage, skill and enterprise.

As soon as I had introduced myself to Mr. Litchfield, I began a carefully prepared inspirational talk. "Mr. Litchfield," I said, "I represent the Earthworm Tractor Company, and I have just discovered a hot prospect for a tractor. Under the old cutthroat methods of doing business, I would have kept this information to myself. But under the New Deal, I feel that we should all work together as a band of brothers. And I am, therefore, inviting you to come along with me."

I then explained all about the airship and the wonderful opportunity it offered for a tractor sale. Mr. Litchfield was very much interested, and he thanked me most cordially for my remarkable generosity in sharing this great opportunity. After acknowledging his thanks in a graceful manner, I explained the difficulties of transporting a tractor to Mr. Mott's island, and outlined my plans for overcoming these difficulties.

"I have made arrangements," I said, "with one of the local hardware and ship-chandler establishments to rent me a half a mile of two-inch rope and a large pulley. Tomorrow morning we will drive our two machines down to the shore at the narrowest point of the channel which separates Mr. Mott's island from the mainland. We will take the pulley and one end of the rope in a dory, and row across to the island. We will fasten the pulley to a tree at the edge of the beach. We will pass the end of the rope through the pulley and bring it back to the mainland in the dory. Do you follow me, Mr. Litchfield?"

"I think so," he replied. "You mean you are going to run this rope from the mainland across the channel to Mr. Mott's island, through a pulley, and then back to the mainland. Is that it?"

"Exactly," I said. "We will then remove from your tractor the magneto, the tools, the seat cushions, and all accessories which might be damaged by salt water. We will plug up the intake and exhaust pipes, the crank-case breather pipe and all openings which might permit water to enter the machine. We will hook one end of the rope to the front of your tractor. The other end of the rope we will fasten to the rear of my machine. I will then drive straight inland, pulling your tractor under water along the bottom of the channel, and over to the island. As soon as this is accomplished, we will row across in the dory, install the magneto and other accessories on your machine, clean out any water which, in spite of our precautions, may have worked into the mechanism and crank up the motor. Then you will pull my machine over to the island. What do you think of the idea?"

"It is most remarkable," said Mr. Litchfield. "Did you think this up all yourself?"

"No, indeed," I said. "It is an old trick in the tractor business. It is tested and tried. Mr. Luke Torkle, one of our oldest and most experienced service mechanics, once told me that he used this method as long ago as 1912 to get two Earthworm tractors across the Congo River in Africa. The same method has been used on various occasions in this country by both tractors and motor trucks. And I have been told that in 1918 a whole fleet of tanks were taken in this way across the Meuse River near Sedan. Of course, it is a rather unusual procedure. But I am sure, Mr. Litchfield, that a man like yourself, who is descended from a long line of venturesome sea captains, would not permit himself to be scared out of an undertaking merely because it is a little unusual."

"It is all right with me," he said. "If you are willing to risk your tractor, I am willing to risk mine."

"Mr. Litchfield," I said, "you are a man after my own heart. I can see that we are destined to get along together fine. It is going to be a real pleasure to cooperate with you. I will now wish you goodbye until tomorrow morning."

With these words, I took my departure and came back to the hotel, where I informed Mr. Jones, the airship man, that I had arranged, for his especial benefit, to have two tractors instead of one. He was, naturally, very much pleased.

In case you have not grasped all of the finer and more subtle points in the arrangements which I am making, I wish to point out that my method of cooperation is so designed as to give us as many advantages as possible. In outlining my plans to Mr. Litchfield, I was careful to state that his tractor would be pulled across the strait first. As he made no objection to this at the time, I am reasonably certain that he will let me proceed in this manner. Although I am confident that my scheme for towing tractors under water will work, I feel that it will be much better to experiment a bit with Mr. Litchfield's machine before trusting our beautiful Earthworm beneath the salt sea waves.

If everything works out all right, and we are successful in our demonstration with the airship, the sale, of course, will be right in the bag for the Earthworm.

Mr. Litchfield's Emperor tractor is a fairly good machine—almost as good as the Earthworm of five years ago. But now that we have our splendid new diesel model with its remarkable performance and low fuel cost, there just is no comparison at all. Mr. Litchfield won't have one chance in a thousand of selling his machine. Thus, I have worked out a scheme of cooperation that is practically perfect—Mr. Litchfield will do half the work, and I will make the sale.

Yours for more and better cooperation,
ALEXANDER BOTTS.

———

ALEXANDER BOTTS
SALES PROMOTION REPRESENTATIVE
EARTHWORM TRACTOR COMPANY

QUITTYQUINK COUNTY JAIL,
CELL NUMBER 2,
QUITTYQUINK COVE, MAINE.
WEDNESDAY AFTERNOON, JUNE 28, 1933.

MR. GILBERT HENDERSON,
SALES MANAGER,
EARTHWORM TRACTOR COMPANY,
EARTHWORM CITY, ILLINOIS.

DEAR HENDERSON: My plans have had a terrific setback, and for the moment I am helpless. But I have not given up hope. Tomorrow morning I expect to be out of jail again, and we shall then see who is to come out on top in this deal.

The unfortunate situation in which I now find myself is due entirely to the fact that my generous and honorable efforts in cooperation were shamelessly taken advantage of by a man whom I considered my friend, but who has turned out to be an unscrupulous and crafty enemy. It is all very distressing, especially in view of the fact that everything seemed so promising this morning when I started my great experiment in the submarine transportation of tractors.

Mr. Litchfield and I arrived with our machines on the shore opposite Mr. Mott's island at about eight a.m., along with several dozen sightseers from the town who had come out on foot and in boats.

The work of running the rope over to the island, passing it through the pulley, bringing it back and then preparing Mr. Litchfield's machine for its underwater journey, took a little more than an hour. And at exactly a quarter past nine, I hooked onto one end of the rope, opened up the motor and slowly but steadily drew the Emperor tractor into the water. Everything worked beautifully, and at twenty-five minutes after nine a spontaneous cheer arose as the machine emerged from the waters and rolled up onto the gravel beach on the other side of the strait.

Mr. Litchfield and I then rowed across to the island and worked for about an hour cleaning the salt water out of every last crack and crevice in the tractor, and installing the magneto and other accessories. At half-past ten we cranked the motor, and it started up with a roar. After Mr. Litchfield and I had congratulated each other on this happy result, I rowed myself back to the mainland, leaving Mr. Litchfield and his tractor on the island. I spent about half an hour in making the Earthworm shipshape for its voyage. I then tied the big rope to the front cross member of the frame and signaled to Mr. Litchfield to go ahead.

I saw him fasten his end of the rope to the drawbar of the Emperor tractor, and then put the machine in motion. The rope tightened, the Earthworm went rolling out into the water, and I hopped into the dory and rowed along right ahead of it.

For a while, everything seemed to go all right. But when the Earthworm reached the middle of the channel—where the water was about twenty feet deep—its forward motion abruptly ceased. The water was so deep that I could not see whether or not it had met some obstacle. In fact, all I could see of the tractor itself was a dim and quivering shadow far down in the watery depths. Looking over toward the island, I noticed that Mr. Litchfield had stopped his machine and got down out of the seat. After waiting for five minutes—during which time nothing happened—I rowed on over to the island, beached the dory, and walked up the shore to the point where Mr. Litchfield was standing beside his tractor. About a dozen of the sightseers from town were gathered around him.

"What's the matter?" I asked. "Has your tractor quit on you?"

"Oh, no," Mr. Litchfield replied. "The old baby is working fine."

"It doesn't seem to be accomplishing much," I said. "I hope my Earthworm hasn't run into a hole, or got so tangled up in the seaweed that you can't pull it."

"No, indeed," he said.

"Then what's the trouble?" I asked. "Why did you stop?"

At this point Mr. Litchfield smiled at me. But it was not a pleasant smile; it was more in the nature of an evil leer.

"There is no trouble at all." he said. "The only thing that has happened is that I have changed my mind about this demonstration."

"What do you mean, you have changed your mind?"

"It has just occurred to me," he said, "that one tractor ought to be enough to pull Mr. Mott's airship into the hangar. If we try to use two, we'll just be getting in each other's way and spoiling the whole demonstration. So, as long as we won't need your Earthworm machine, there is no reason for bringing it over here. It will be much better to leave it in its present location—where it is perfectly safe—until after the airship has landed. As soon as the demonstration is over, we will start figuring on how to tow it back to the mainland."

As he spoke these words, Mr. Litchfield smiled even more broadly than before. And never in my life have I seen a more obnoxious and repulsive smirk.

Now, I am not by nature a violent man. As well you know, Henderson, I have always been noted for my sweetness of disposition and my magnificent self-control in the face of all the usual annoyances which form a part of the life of a tractor salesman. But the fiendish ingenuity and monumental effrontery of this poisonous reptile were a little too much even for me. For the moment, I fear I lost control of my temper. But this did not in any way interfere with my sense of timing, or with my muscular coordination. After a brief wind-up, I made a long and graceful swing, and landed my right fist, with the entire weight of my body behind it, just below Mr. Litchfield's left eye.

A theoretical moralist might argue that I should not have acted in this hasty and brutal manner. But, on the whole, I am glad that I did so. Even now—several hours after the event—when I recall the dull but magnificent thud with which my knuckles came in contact with that evil visage, I am filled with a feeling of indescribable satisfaction.

However, there have been certain other consequences of my act which are not so pleasing. In the excitement, it had momentarily slipped my mind that Mr. Litchfield was the county sheriff. He was not long in reminding me of

"The Earthworm went rolling
out into the water."

this fact however. After staggering about for a moment or two, he began yelling for help. A half a dozen of the bystanders grabbed me before I could get in a second blow, and Mr. Litchfield informed me that I was under arrest for assault and battery, attempted murder and resisting a peace officer.

Two of the men who were holding me turned out to be deputy sheriffs. Acting under Mr. Litchfield's orders, they marched me down to the shore, loaded me into a motorboat, brought me back here to town and locked me in the jail— where I am still languishing, and where, they tell me, I will have to continue to languish until I am taken before the justice of the peace at nine o'clock tomorrow morning.

The jail is modern and sanitary, and from my window I have a very pleasing view, looking out over the ocean and Mr. Mott's island. I am comfortable enough, but it is, naturally, most annoying to have to sit around doing nothing when I ought to be working hard on the difficult problem of getting that tractor out of the ocean and preparing it for tomorrow's demonstration.

I am not licked yet however. As soon as I explain the situation to the justice of the peace tomorrow, he will see that my affair with Mr. Litchfield comes under the classification of justifiable, and even praiseworthy, assault. He will discharge me, probably with the thanks of the court. I will then start rushing around to see if I can work out some scheme for getting my tractor the rest of the way to the island. If I cannot accomplish this before the airship arrives, I will be on hand myself anyway, and I will be prepared to give Mr. Mott a description of the advantages of our diesel Earthworm tractor that will just naturally compel him to start reaching for his fountain pen and looking for the dotted line.

I can produce a sales talk that will be better than anybody else's sales talk and demonstration combined. It will be especially easy this time, because I am always more eloquent when I am mad. And right now I am just about as mad as I ever get.

This whole affair just goes to show that you cannot trust anybody who is descended from a long line of New England sea captains. It is a matter of common knowledge that these Yankee skippers used to sail all over the world, swindling each and every man with whom they did business, and giving to the name "American" an unsavory reputation in every land which they visited. With such ancestry, it is no wonder that Mr. Litchfield is about as low as they make them. As a matter of fact, I was suspicious of him from the first. Whenever you meet a man who shakes hands in an unnecessarily hearty and virile manner, and who goes out of his way to look you straight in the eye, you can be sure he is doing it to cover up a very ripe state of inward corruption. A final touch is added by the fact that Mr. Litchfield is county sheriff—a sure indication that, in addition to his other vices, he goes in for dirty politics.

"After a brief wind-up, I made a long and graceful swing."

In the past he has apparently been successful in his nefarious schemes. But in locking horns with me, he will find that he has picked up a bigger cat than he can swing. Just wait until I get out tomorrow.

LATER, NEXT MORNING,
THURSDAY, 10 A.M. JUNE 29, 1933.

What a world this is! When a man is down, everybody starts to kick him. As I wrote you yesterday afternoon, I confidently expected to get out of jail this morning. A little while ago, however, one of the deputies came in and said the justice of the peace had decided to postpone my hearing until tomorrow. The old duffer was so afraid that he might miss the thrilling spectacle of the landing of the dirigible—scheduled for this afternoon—that he went over to the island at about five a.m. He is going to camp over there until the airship appears. This, at least, is the explanation given me by the deputy. But the whole thing smacks of the Machiavellian hand of Mr. Ezra Litchfield, the dirty skunk. He has probably brought his slimy influence to bear upon this justice of the peace in order to keep me out of the way and give him a clear field for his own demonstration and sales talk.

The whole thing is an outrage, but I have not been able, so far, to do anything about it. Last night I talked on the telephone with Mr. Jones, the airship man, and begged him to help me. But he claimed he could do nothing, and seemed to take very little interest in my case. All he cares about is getting his airship landed properly. He is going to use Mr. Litchfield's tractor, and he forgets entirely that I am the guy that thought of the whole idea in the first place. As I could get nothing out of him, I called our local Earthworm dealer, but all I could get out of him was a request to stop bothering him while he was sick, and a statement that he intended to sue me for losing his tractor in the bottom of the ocean.

This is a bad business. It is bad for me, but it is even worse for Mr. Mott. He will see the Emperor tractor pulling the dirigible into the shed, and after listening to Mr. Litchfield's guileful and misleading sales talk, he will probably buy it. Naturally, I feel the deepest sympathy for poor Mr. Mott. If I could help him, I would do so. But just at the moment I seem to be pretty well blocked.

LATER. 3 P.M.

I have just been watching, from the window of this admirably located jail, the landing of the great all-metal airship. The old baby came gliding in out of the mists of the Atlantic, circled once over the island and then settled gracefully down onto the landing field, its aluminum sides gleaming in the afternoon sunshine. It was a truly inspiring and magnificent sight, but I was in no proper state of mind to appreciate it.

I borrowed some field glasses from the deputy in charge of the jail—I am the kind of guy that even when I am in jail, I make friends with the jailer—and with these glasses I was able to see the details of the landing very clearly. I could see a line thrown down from the bow of the airship and fastened to the Emperor tractor. Another line was thrown down from the stern of the ship, and this should have been fastened to my tractor. But, as my tractor was not there, it was grabbed by a crowd of bystanders. Probably Mr. Jones had hired them for this purpose. I saw the tractor and the people start across the field, with the great ship floating along above them. There was practically no wind. Slowly and majestically the ship moved to the open doors of the hangar, and then disappeared inside. The landing was perfect.

I don't know whether Mr. Mott has decided to buy the airship. If he has, it is very probable that he is, at the very moment, arranging with the nefarious Mr. Litchfield for the purchase of one or more of those unspeakable Emperor tractors. And here I am, with my tractor at the bottom of the ocean, and myself locked up in jail.

The situation is enough to drive a man insane. If I ever have to go through another afternoon like this, they will have to put me in an asylum rather than a jail.

But there will be no more such afternoons in my life. From now on I am through with cooperation. Never again, at any time, at any place, will I ever, for any reason, cooperate, or even attempt to cooperate, in any way, with any person, in any manner whatsoever. I have learned my lesson.

QUITTYQUINK HOTEL,
QUITTYQUINK COVE, MAINE.
THE NEXT DAY, FRIDAY NOON.

Well, I am out of jail again—having been turned loose about nine o'clock last evening.

The circumstances surrounding my release were, on the whole, rather remarkable. I had just about made up my mind that I would have to spend another night in the lock-up, and I was just starting to get undressed preparatory to going to bed. As I was untying my shoes, humming gently the melancholy strains of "The Prisoner's Song," two visitors arrived— Mr. Jones, of the airship company, and Mr. Mott, the big millionaire.

Mr. Jones was all in a twitter, announcing loudly that Mr. Mott had decided to buy the airship, and wasn't it wonderful? He seemed to be highly enthusiastic—as well he might. Mr. Mott, who was a very intelligent and forceful-looking person, seemed to be calm and unconcerned. My own condition might be described as morose.

Mr. Jones explained that Mr. Mott, having observed the beautiful black eye which, it appears, is now a most conspicuous feature of Mr. Ezra Litchfield's face, had expressed a desire to see the man who was responsible for it.

"So that's it," I said. "He comes to peer at me through the bars of my cage as if I were some strange animal at the zoo."

"Don't get insulted," said Mr. Jones. "He also wants to talk business with you."

"Yes," said Mr. Mott. "I will need two tractors to handle my dirigible here. And, after thinking over the interesting demonstration which you and Mr. Litchfield put on, I have decided that I want Earthworm machines."

"Say, that's swell!" I said. "But I don't understand about the demonstration. I wasn't there, and the Earthworm tractor wasn't there. You haven't even seen my machine, unless you caught a glimpse of it down under twenty feet of water as you came over from the island."

"I'm not talking about the machines," said Mr. Mott. "I'm talking about you and Mr. Litchfield personally."

"But what have we got to do with it personally?"

"Everything," said Mr. Mott. "Mr. Jones here has seen both your tractors, and he tells me that either make is good enough to handle the airship. As far as the machines themselves are concerned, one would be as satisfactory to me as the other."

"Oh, no, it wouldn't," I said. "If you had seen the Earthworm in action—"

"It wouldn't have made any difference," Mr. Mott interrupted. "You see, Mr. Jones has told me all about the way you got Mr. Litchfield's tractor to the island, and what happened afterward. And that, to me, is the most important part of the demonstration."

"You mean you like to deal with a company that has bright men working for it, and you were impressed by the clever way I got that big tractor across the channel?"

"No," said Mr. Mott. "I was impressed by the clever way in which Mr. Litchfield stole the demonstration away from you and showed you up for a sucker by leaving your tractor way out there under water."

"You mean you are going to buy my machine because you are sorry for me?"

"Oh, no. If I buy a couple of tractors, it is probable that I will have to have a certain amount of service on them—repair work, new parts and so on. That's right, isn't it?"

"Yes," I said.

"In other words," he went on, "as long as I own these machines, I will have to have business relations with the man that represents the company which made them. If I bought the Emperor machine, I would have to deal with Mr. Litchfield. And he is far too clever for me. I much prefer to deal with a man like yourself, who is credulous, helpful, always ready to cooperate with the other fellow, and more or less gullible. You would not be smart enough to slip anything over on me. In fact, I could probably slip quite a bit over on you. And that is why I am buying the Earthworm, instead of the Emperor."

"Mr. Mott," I said, "your reasoning may be cockeyed, but your conclusion is sound. Your words fill me with joy. Now, as soon as I can get out of this filthy jail and figure out some method of rescuing my tractor from the ocean—"

"That has been all arranged," said Mr. Mott, "through the cooperation of Mr. Litchfield."

"You don't mean that Mr. Litchfield is actually going in for real cooperation?"

"He has to," said Mr. Mott. "He has to get that tractor of his back to the mainland. I told him that if he would haul your tractor the rest of the way to the island, and also drop the charges against you and release you from jail, I was sure you would be willing to pull his tractor across the channel. Incidentally, I agreed to pay him a fair amount for his time and trouble in helping me land my dirigible. At first, he was pretty sore because I would not buy an Emperor tractor, but in the end he agreed to my entire proposition. Before we left the island he pulled your machine out of the water. It seems to be in perfectly good condition, so I will accept it as one of the machines which I am buying from you. On the whole, Mr. Litchfield is showing a very commendable spirit in this thing. He seems to be a good sport and a good loser."

"He ought to be," I said. "He is descended from a long line of New England sea captains—sturdy men, who had to learn how to accept success or failure, all as part of the day's work. It is natural that a descendant of such people should possess the same courage."

"Apparently, then," said Mr. Mott, "you hold no grudge against this gentleman. I take it you will be perfectly willing to pull his machine across the channel for him?"

"I should be delighted," I said. "I am always glad, at any time, to cooperate with anyone in any manner whatsoever."

"Splendid," said Mr. Mott. He then called in Mr. Litchfield, who had been waiting downstairs, and I was promptly let out of jail.

This morning I cleaned up the tractor and pulled Mr. Litchfield's machine back to the mainland. You will be glad to know that during all the time I was working with Mr. Litchfield I treated him with the most effusive politeness and consideration. This showed him what a kind and Christian disposition I have, and also had the happy result of making him madder than almost anything else I could have done.

I am now back on the mainland at the hotel. An order for one additional Earthworm is going forward through the office of the local agency. When this machine arrives, it will be taken to the island by my highly successful method of underwater transportation.

In the meantime, I will stick around here and see if I can stir up a few prospective tractor buyers in the lumber business. It is even possible that I may be able to entice Mr. Litchfield into some new cooperative scheme.

Very sincerely,
ALEXANDER BOTTS.

THE GREAT HIGHWAY CONTROVERSY

ILLUSTRATED BY TONY SARG

EARTHWORM TRACTOR COMPANY
EARTHWORM CITY, ILLINOIS
OFFICE OF THE SALES MANAGER

SATURDAY, JULY 8, 1933.

MR. ALEXANDER BOTTS,
BAXTER HOTEL,
BAXTER HARBOR, MAINE.

DEAR BOTTS: You will proceed at once to Belfair, Maine—where, at the moment, we have no dealer—and you will get in touch with the Belcher Lumber Company. The president of this concern has just written us that he expects, in the near future, to purchase five large tractors for logging operations.

If you hurry, you ought to be able to land this business for the Earthworm Tractor Company.

Very sincerely,
GILBERT HENDERSON,
Sales Manager.

—————

ALEXANDER BOTTS
SALES PROMOTION REPRESENTATIVE
EARTHWORM TRACTOR COMPANY

BAXTER HOTEL,
BAXTER HARBOR, MAINE.
TUESDAY EVENING, JULY 11, 1933.

MR. GILBERT HENDERSON,
SALES MANAGER,
EARTHWORM TRACTOR COMPANY,
EARTHWORM CITY, ILLINOIS.

DEAR HENDERSON: Your letter of July eighth has come, but I am so busy I have hardly had time to even read it.

I am still at Baxter Harbor. As the local dealer is sick, I am running his business for him. And you will be delighted to hear that I have just sold—

or practically sold—ten of our new-model, seventy-five-horsepower, diesel tractors. The whole thing is as sudden and unexpected as it is delightful. This morning the local paper published a statement by the state highway department, saying that they are planning to build a magnificent new highway—or motor parkway, as they call it—from Baxter Harbor here to the town of Indian Bluffs, about twenty-five miles up the coast. The new road will be cut straight through what is known as the Baxter Forest—a tract of rough and densely wooded land about five miles wide, and extending along the seashore all the way from Baxter Harbor to Indian Bluffs.

As soon as I had read about this splendid project—which means lots of dirt moving and grading, and hence lots of work for tractors—I went around and called on one of the state-highway engineers, a man by the name of Dudley, who is here in town to boss the job. Mr. Dudley is rather quiet and studious-looking, but I feel that he is a good egg. Although he did not definitely commit himself, he hinted very strongly that when they started work they would need no less than ten large tractors. He also said that he considered the Earthworm the best machine for this work. So, you see, the sale is practically in the bag.

Yours,
ALEXANDER BOTTS.

P.S. I forgot to say that there is one minor feature in the situation which may interfere with our selling these tractors. The state highway commission, although it has worked out plans for this new road, has not, as yet, definitely decided to build it. It seems that state and Federal funds are available to meet the cost of construction, but before they go ahead they want to be sure that the people of the region here really want this improvement. The announcement in the paper this morning contained a complete description of the proposed highway, and stated that a public hearing is to be held on Tuesday evening, July twenty-fifth. If popular opinion is favorable, the construction will start at once. If it is unfavorable, the matter will be dropped. But I have no fears. This is a swell project, and everybody will be for it.

Very truly,
ALEXANDER BOTTS.

P.S. Another possible obstacle to our selling these tractors is the fact that Mr. Abner Smith, the local dealer for the Emperor tractor, is making strenuous efforts to capture this business. When I arrived at Mr. Dudley's office this morning, I met Mr. Smith coming out. I was pleased to observe, however, that he looked pretty worried and discouraged. I shouldn't wonder if Mr. Dudley had told him perfectly frankly that his wretchedly constructed Emperor machine had no chance at all on this magnificent highway project. Of course, Mr. Smith will buzz around and do everything he can to make a sale. But I am not worried. With me here to present the advantages of the Earthworm, he is licked before he starts.

Yours,

ALEXANDER BOTTS.

———

TELEGRAM

EARTHWORM CITY ILL JULY 14 1933
ALEXANDER BOTTS
BAXTER HOTEL
BAXTER HARBOR ME

YOUR LETTER SAYS NOTHING ABOUT GOING TO BELFAIR STOP IF YOU HAVE NOT ALREADY DONE SO YOU WILL GO AT ONCE STOP OTHERWISE WE MAY LOSE IMPORTANT ORDER FROM BELCHER LUMBER COMPANY STOP IF YOU THINK ADVISABLE YOU MAY ARRANGE TO RETURN TO BAXTER HARBOR TO COMPLETE DEAL WITH STATE HIGHWAY DEPARTMENT

GILBERT HENDERSON

ALEXANDER BOTTS
SALES PROMOTION REPRESENTATIVE
EARTHWORM TRACTOR COMPANY

BAXTER HOTEL,
BAXTER HARBOR, MAINE.
FRIDAY EVENING, JULY 14, 1933.

MR. GILBERT HENDERSON,
SALES MANAGER,
EARTHWORM TRACTOR COMPANY,
EARTHWORM CITY, ILLINOIS.

DEAR HENDERSON: Your telegram is received, and I will get down to Belfair just as soon as I can. At the moment, however, my presence is urgently needed here to combat an unusually lousy trick which has just been pulled on me by Mr. Abner Smith, the local dealer for the Emperor tractor.

Last Tuesday, as I related in my letter of that date, this guy was making strenuous efforts to horn in on my prospective sale to the State Highway Department. On Wednesday, owing to the fact that he was away somewhere on a business trip, he was forced to interrupt his activities. But on Thursday—yesterday—he got back, and immediately started to act like a spoiled child. Having finally recognized the fact—which he should have known from the beginning—that he had absolutely no chance to sell his pathetically inefficient Emperor tractors in competition with our Earthworms, he allowed his entire mind to become permeated with envy and sullen hatred. He decided that if he could not get this business, nobody else could either. And as a result of his dog-in-the-manger attitude, he has initiated a malicious, but very plausible, attack on this whole highway project.

In the furtherance of his nefarious scheme, he spent most of yesterday in contacting various members of a local organization known as The Baxter Trail Club. This club is composed of nature lovers who are interested in the Baxter Forest. And although it does not own the land, it has, in a very public-spirited manner, constructed a trail from one end of this woodland to the other, and built small overnight cabins every few miles for the accommodation of anyone who wants to get out into the wilderness.

Mr. Abner Smith, by working upon the natural pride which the club members have in their trail, has succeeded in convincing them that a motor highway through the forest would completely ruin the usefulness of this dinky little footpath. In fact, he has been so successful with his falsehoods, misrepresentations and distortions of truth that he has succeeded in getting these good people worked up into a state of almost frenzied excitement and indignation. Not only are they opposing the highway by word of mouth among their friends but they have started bursting into print.

When I opened the paper this morning, the first thing I noticed was a rather plaintive letter from the president of The Baxter Trail Club, who, it appears, is a well-meaning but doddering old gentleman by the name of Quackenbush. Mr. Quackenbush, after dwelling at great length on the unspoiled beauty of the Baxter Forest, the "cool dark shade of the giant trees," and "the spiritual peace which comes only through close communion with nature," expressed the pious hope that this "primeval sanctuary" might be spared from the desecration of a motor road.

Mr. Quackenbush's remarks were pretty weak and sentimental, but they made me nervous. When a man has a sale of ten tractors hanging in the balance, it doesn't take much to upset him. As I read on through no less than six similar letters from other members of the club, my nervousness increased. And when I finally came to a communication from the arch conspirator, Mr. Abner Smith, himself, I was seriously alarmed. In order that you may understand my feelings, I will quote the letter in full.

"I have just been informed," wrote Mr. Smith, "that some demented engineer is actually proposing to build a vile motor highway through the very heart of the Baxter Forest. This vast wilderness—this peaceful refuge from all the nerve-racking noise and hurry of our modern machine-made civilization—this holy temple of Mother Nature—is to be 'improved' by a vast strip of concrete pavement so that speed-maddened motorists may drive their roaring machines through glades which once knew no sound but the delicate footfall of the deer and the sweet piping of the wood thrush. The very air of the forest, now heavy with the scent of pines, is to be befouled with the noisome stench of thousands of exhaust pipes, while the peace of the ages will be rent by the incessant honking of motor horns. And, worst of all, every hillside will blossom forth with unspeakable billboards and advertising signs. Every valley will sprout a poisonous, mushroom growth of miserable tourist cabins. And the entire landscape will be dotted with sordid hot-dog stands and mephitic filling stations.

"The colossal effrontery of this ghoulish scheme is enough to make one's blood boil. It is a nightmare of commercial sordidness. And every right-thinking man and woman in the entire state of Maine should rise up in active opposition to the barbarous marauders whose one purpose is the destruction of all that is fair and beautiful."

When I had finished reading the above letter, I was—as I have already stated—considerably alarmed. My nervousness in regard to the ten tractors I am hoping to sell became so pronounced as to be almost painful. Mr. Abner Smith's arguments, of course, were completely cockeyed, but they were expressed so strongly that I was afraid a certain number of people might take them seriously. In order to check up on the state of public opinion, I made a tour of the local garages and filling stations, which in recent years have largely taken the place of the old general stores as forums for the discussion by the natives of all matters of general interest.

The results of this tour—which occupied most of the day—were not reassuring. Almost everybody seemed to sympathize with Mr. Quackenbush, who apparently is a rather nice old duffer and is very highly regarded in the community. In addition, I heard many expressions of admiration for the snappy way in which Mr. Abner Smith handles the English language. There seemed to be a very definite trend of public opinion opposing the new road.

"I spent last Saturday on a tour of all the local filling stations."

Having ascertained this fact, I called late this afternoon on Mr. Dudley, the state engineer who is in charge of the proposed highway project. He seemed surprised and distressed at the sudden hostility to his plans. But when I asked him if he was making any effort to rally public sentiment in our favor he replied, rather weakly, that he had not yet decided what action he ought to take.

"The public doesn't seem to understand what we are trying to do," he said, "in spite of the fact that we have already published a complete explanation of our plans. The only thing I can think of would be to explain the whole thing over again. But if they didn't understand it the first time, they probably wouldn't the second time."

"Mr. Dudley," I replied, "you may be a good engineer. But as a molder of public opinion, you are rotten. Something, however, must be done. And as you are doing nothing, I will have to appoint myself director of publicity for your splendid project. I will now sally forth and start a counterattack against our enemies. Good day."

And before he could even thank me, I rushed away and came back to the inn here, where I have been planning out a vigorous and powerful publicity campaign. Tomorrow morning I will go into action with all the energy I possess, and by the time I get through with my opponents they will be just laying about in heaps.

Yours,

ALEXANDER BOTTS.

————

TELEGRAM
EARTHWORM CITY ILL JULY 17 1933
ALEXANDER BOTTS
BAXTER HOTEL
BAXTER HARBOR ME

HAVE YOU BEEN TO BELFAIR YET QUESTION MARK

GILBERT HENDERSON

ALEXANDER BOTTS
SALES PROMOTION REPRESENTATIVE
EARTHWORM TRACTOR COMPANY

BAXTER HOTEL,
BAXTER HARBOR, MAINE.
MONDAY EVENING, JULY 17, 1933.

MR. GILBERT HENDERSON,
SALES MANAGER,
EARTHWORM TRACTOR COMPANY,
EARTHWORM CITY, ILLINOIS.

DEAR HENDERSON: In reply to your telegram, which has just been received, I beg to advise that I have been unable, as yet, to go to Belfair. But don't worry. I will rush down as soon as I possibly can—probably about the end of next week. In the meantime, I am working early and late on my publicity campaign against the opponents of the great new highway. This campaign is based on the principle that the best defense is a vigorous attack.

In planning my offensive, I have entirely disregarded the published letters of Mr. Quackenbush and the other members of The Baxter Trail Club. The complaints of these gentlemen, being mere sentimental tripe, offer very little opportunity for brilliant rebuttal. I am, therefore, concentrating my artillery on a vulnerable point which my quick mind has discovered in the arguments presented in the letter which Mr. Abner Smith sent to the paper. In this letter Mr. Smith was incautious enough to speak sneeringly, not to say unpatriotically, about those sturdy and well-established American institutions, the billboard, the filling station, the hot-dog stand and the tourist cabin. And I am now capitalizing on this mistake of his in a big way.

I spent last Saturday morning on a whirlwind tour of all the local filling stations—seven in number. I carried with me a copy of the paper containing Mr. Smith's letter, and I also carried a copy of an unabridged dictionary. The dictionary was very heavy, but I was well repaid for my efforts in dragging it around. In interviewing each filling-station proprietor, I would first show him Mr. Smith's allusion to *mephitic* filling stations. I would then open the dictionary, and point to the definition of *mephitic*—"poisonous; pestilential; foul; noxious; skunk-like." In every case this got a good reaction. And by noon I had organized all the filling-station proprietors into a protective trade association; I had written out and persuaded them to sign a statement for the paper which contained a violent denunciation of Mr. Smith's gratuitous insult to an honest and useful business, and expressed their enthusiastic approval of the new high-way on the ground that it would increase local prosperity.

In the afternoon, I succeeded in organizing the owners of various tourist cabins and refreshment stands, and prevailed upon them to sign additional statements in favor of the proposed road.

All of these statements, owing to the fact that I wrote them myself, were remarkable for their vigor and aggressiveness.

Having completed my work with these business men, I spent all of Sunday in contacting representatives of the general public. By Sunday night, I had persuaded fourteen different people to send letters to the paper. All of them were published this morning—Monday—and some of them were swell. I found one old codger who is a connoisseur of hot dogs. Whenever he takes a trip his main enjoyment comes from sampling the different varieties which he finds at the stands along the road. He stated that he is heartily in favor of the new highway because it is sure to mean more and better hot-dog stands.

One of the waitresses at the hotel here, to whom I had given an unusually large tip, was obliging enough to send in a letter in praise of billboards and advertising signs. She says that the handsome and distin-guished young ladies and gentlemen who are so attractively portrayed in the large-sized cigarette advertisements provide her with invaluable and up-to-the-minute ideas on chic sports attire, as well as more formal dress. She also says that she took a motor trip to California last year. She reports that the Yosemite Valley is nothing but rocks and silly waterfalls. The Yellowstone is incredibly dull, and the great plains are perfectly poisonous. The one redeeming feature of the trip was the intellectual pleasure she derived from the witty wise cracks on the signs put up by the Balmy Shaving Cream Company. She is highly enthusiastic over the new highway. What this part of the country needs, according to her, is more good advertising signs.

An elderly couple who drives down to Florida every winter sent in a particularly good letter, explaining that they are too old to sleep outdoors, have insufficient funds to stay at city hotels and would be deprived of their winter vacation if it were not for the cheap tourist shacks which they find along the way.

"And as a grand finale he recited, in a voice...

The snappiest idea of all was a letter I wrote myself, and signed with a fictitious name. As a matter of fact, the letter was fictitious, too, but the editor of the paper never got wise, and he published it along with the rest. It read as follows:

"I am nothing but a cripple—a poor old man with one leg—but I wish to add my humble plea to the discussion of the proposed highway. For years I have been hearing of the beauties of the Baxter Forest. For years my greatest desire has been to visit this wonderful primeval wilderness. I have particularly wanted to see the great hickory tree from which, long ago, my brother cut a limb to fashion into a wooden leg for me. But, even with this leg, I am too lame to walk out over the trail. My only hope for realizing my dearest wish lies in the construction of this proposed highway. If the opponents of the plan succeed in blocking it, I want them to know that they will also succeed in breaking the heart of a poor old cripple."

I am particularly pleased with this last letter, owing to the fact that it injects a tender sentimental note into a controversy which has, of necessity, contained a good deal of bitterness and sarcasm.

As a result of my efforts, it begins to look as if the great tide of popular opinion against the highway has been definitely stemmed. But we have not yet won the day. So I am keeping on with the good work in my usual relentless fashion. In addition to thinking up more letters for the paper, I am conducting a vigorous word-of-mouth campaign.

This morning I asked the president of the Baxter Harbor Rotary Club for permission to address their weekly meeting this noon. He stated that they already had a speaker scheduled, but invited me to attend—which I did. You can imagine my disgust when I found that the speaker was Mr. Abner Smith, himself. It appears that he is quite an effective orator in a cheap sort of way. And he is also pretty good at reciting poetry—they say he is quite famous around these parts for his rendition of "The Shooting of Dan McGrew," and "Gunga Din." At the meeting this noon he delivered a very emotional address on the subject of the proposed highway, which he referred to as "an attempted rape of the virgin loveliness of the wilderness." He had a lot to say about protecting "our little feathered songsters"—meaning birds, I suppose—and the "wee, furry, wild things"—referring probably to musk-rats, skunks and other small mammals. And as a grand finale he recited, in a voice choked with emotion, a poem called "Trees." When he sat down he had all those hard-boiled business men pretty near in tears. And I will have to admit that as I observed the effect this guy was making, and as I meditated on those ten tractors which seemed to be getting farther and farther from an actual sale, I was almost ready to weep myself.

...choked with emotion, a poem called 'Trees.'"

But don't get the idea that I am discouraged. I am keeping after this thing like a professional bird lover running down a new species of feathered songster. And I want you to know that I am going to stay here on the job right up until the public hearing next Tuesday evening.

Sincerely yours,
ALEXANDER BOTTS.

———

TELEGRAM
EARTHWORM CITY ILL JULY 20 1933
ALEXANDER BOTTS
BAXTER HOTEL
BAXTER HARBOR ME

WE DISAPPROVE COMPLETELY OF YOUR RECENT PROPAGANDA ACTIVITIES STOP WE WOULD RATHER LOSE THE SALE OF TEN TRACTORS THAN COMPROMISE THE HIGH REPUTATION OF THIS COMPANY BY SUPPORTING A PROJECT WHICH AIMS TO DESTROY THE NATURAL BEAUTY OF THE COUNTRY STOP THE FACT THAT THE REPRESENTATIVE OF THE EMPEROR TRACTOR IS ACTING IN AN UNSELFISH AND PUBLIC SPIRITED MANNER MAKES IT DOUBLY IMPORTANT FOR US TO AVOID ANTAGONIZING THE MORE THOUGHTFUL MEMBERS OF THE COMMUNITY STOP YOU WILL DISCONTINUE YOUR CLOWNISH CAMPAIGN ON BEHALF OF HOT DOG STANDS BILLBOARDS AND ALL THE EYESORES WHICH UNFORTUNATELY CLUTTER UP SO MANY OF OUR AMERICAN HIGHWAYS AND YOU WILL GO TO BELFAIR AT ONCE AND CALL ON THE PRESIDENT OF THE BELCHER LUMBER COMPANY STOP IT IS IMPERATIVE THAT YOU GO AT ONCE TO AVOID LOSING IMPORTANT SALE

GILBERT HENDERSON

ALEXANDER BOTTS
SALES PROMOTION REPRESENTATIVE
EARTHWORM TRACTOR COMPANY

BAXTER HOTEL,
BAXTER HARBOR, MAINE.
THURSDAY EVENING, JULY 20, 1933.

MR. GILBERT HENDERSON,
SALES MANAGER,
EARTHWORM TRACTOR COMPANY,
EARTHWORM CITY, ILLINOIS.

DEAR HENDERSON: Your splendid long telegram has come. I have just been counting up, and I find that it has one hundred and thirty-four words, not counting the address and signature. It must have cost quite a bit to send all those words from Illinois to Maine, but it is well worth the expense. I feel that you have hit the nail right on the head, and I want you to know that I agree with you absolutely. You are perfectly right in your feeling that my campaign in favor of billboards and hot-dog stands was ill-advised. As a matter of fact, I had already come to the same conclusion all by myself.

This change of front on my part was brought about yesterday morning at an interview with Mr. Dudley, the engineer who is promoting the new highway. I had called on him to explain that I was the instigator of the great flood of letters to the paper in favor of his plan, and to receive the thanks which I supposed he would give me for my efforts in his behalf. Much to my surprise, however, he seemed far more irritated than pleased.

"I wish," he said, "that you people who engage in public controversies would take the trouble to familiarize yourselves with the facts of the case. I do not believe you even read my public statement on this new highway project."

"Well, Mr. Dudley," I said, "it was pretty long and dull, so perhaps I did only skim through it. But I got the main points."

"It does not seem to me," he said, "that the main points have been grasped by any of the people who have been rushing into print, either for or against my plan. All the discussion of hot-dog stands and billboards is aside from the point. There will be no such things on my proposed highway."

"What?" I said. "A state road without hot-dog stands and billboards? I never heard of such a thing."

"You have never heard of the difference between a parkway and an ordinary highway?"

"No, I guess not."

"In my statement to the press," said Mr. Dudley, somewhat feelingly, "which you did not even take the trouble to read, I explained that in the case of an ordinary highway the land along the sides of the road is privately owned. But in the case of a parkway it is publicly owned."

"What difference does that make?" I asked.

"All the difference in the world. The more you improve an ordinary highway, the worse it gets. The more it becomes cluttered with roadside nuisances, the more it acquires traffic hazards—crossroads and private driveways from which cars come shooting out into the main stream of traffic. But the more you improve a parkway, the better it gets. It is permanently protected from all undesirable developments."

"It sounds like a good theory," I said. "I wonder if it could be worked out in practice."

"It has been worked out already," said Mr. Dudley, "in a few roads like the Bronx River Parkway north of New York City. But we do not have enough roads of this kind. What this country needs is a complete national system of parkways to handle all through traffic, leaving the present highways for local traffic."

"You're not thinking of building this whole system right away, are you?"

"No, but I had hoped to make a start here in Maine. Our plan for the Baxter highway calls for the purchase of the whole Baxter Forest—five miles wide—as a state park. This would give more than adequate protection to the roadsides. We would have a real parkway, and as soon as people saw how good it was, they would want more. If we can only put this project through, I am sure it would start a much-needed movement for more intelligently planned highways."

"Mr. Dudley," I said, "you don't know how you impress me. Your ideas are so much better than I had supposed that I am lost in admiration at your intelligence. And I am ashamed of myself for not reading with sufficient care your statement in the paper. It is most fortunate that I came in to see you. In any controversy the people who do the talking are, of course, the most important. But it is, nevertheless, highly desirable that there be someone in the background who knows the basic facts and understands what it is all about. As a result of this interview, I am going to change my whole plan of campaign. And the change will be advantageous."

"Almost any change in your campaign would probably be a good thing," said Mr. Dudley. "But are you sure that even now you understand what it is all about? Don't you think it would be best if you were just to drop out of the controversy entirely?"

"Absolutely not," I said.

"But exactly what are you planning to do next?" he asked.

"I do not know," I said. "I am so bewildered by all these new ideas which you have given me that I will have to go back to my room and collect my thoughts. But you may be sure that I will do something. Whatever it is, it will be both sensational and effective."

Without giving Mr. Dudley time for anymore protests, I took my departure and returned here to the hotel. After a prolonged period of meditation yesterday afternoon, last night and this morning, I have finally worked out a plan for my future activities.

I have resolved, for the present, to do nothing toward removing the barnacles of misinformation which have incrusted this controversy. I will let these boobs continue their great hot-dog-stand argument—which has nothing to do with the case—until the latter part of the public hearing on Tuesday. By this Fabian policy I will put myself in a very strong position indeed. I will be the only speaker at the meeting who will know what he is talking about. Mr. Dudley, of course, doesn't count; he knows his stuff, but he is no speaker.

The meeting is sure to be a lallapaloosa. I will let one opposition speaker after another holler and roar about the hot-dog stands and the billboards. The more they deepen the fog of ignorance and misunderstanding, the more they will help set the stage for me. At the proper dramatic moment, I will arise and deliver an address so weighted down with cold facts and so lifted up with warm emotion that it will act as a great searchlight of truth cutting through the murky shadows of falsehood. The entire hot-dog-stand argument will fall flat. Our opponents will be shown up as the saps which they are, and the meeting will be stampeded into a great ovation in favor of Mr. Dudley's wonderful parkway.

I have already started to write up my big speech. As long as Mr. Dudley's plan will discourage roadside nuisances and protect the wilderness by buying up the forest as a public park, I will have much to say about the wee, furry, forest folk and our little feathered songsters. And the scathing remarks which I will make about hot-dog stands and billboards will practically burn the paint off the inside of the building. This, of course, will be an exact reversal of my previous position in reference to hot-dog stands. But that doesn't bother me at all, because I am completely convinced that this time I am right. As a matter of fact, I have always disliked billboards and roadside shacks. But my point of view was temporarily distorted by the fact that I dislike even more to lose the sale of ten tractors.

Between now and the meeting on Tuesday evening I will stay right here, working on my address and keeping my ear to the ground in such a way as to be in constant touch with all phases of public sentiment.

Very truly,
ALEXANDER BOTTS.

———

TELEGRAM
EARTHWORM CITY ILL JULY 24 1933
ALEXANDER BOTTS
BAXTER HOTEL
BAXTER HARBOR ME

YOUR LETTER RECEIVED STOP YOU WILL TAKE YOUR EAR OFF THE GROUND AND GET DOWN TO BELFAIR AT ONCE STOP FAILURE TO OBEY THIS ORDER WILL RESULT IN YOUR IMMEDIATE DISMISSAL
GILBERT HENDERSON

ALEXANDER BOTTS
SALES PROMOTION REPRESENTATIVE
EARTHWORM TRACTOR COMPANY

BAXTER HOTEL,
BAXTER HARBOR, MAINE.
WEDNESDAY EVENING, JULY 26, 1933.

MR. GILBERT HENDERSON,
SALES MANAGER,
EARTHWORM TRACTOR COMPANY,
EARTHWORM CITY, ILLINOIS.

DEAR HENDERSON: Many thanks for your telegram, which arrived day before yesterday—Monday—afternoon. From the wording of your message, I gathered that you were getting quite nervous over my delay in going to Belfair. And as I had finished writing my big speech, and had nothing really pressing until the Tuesday-night meeting, I decided I might as well follow your suggestion. Accordingly, I caught the night train to Belfair, and bright and early the next morning—yesterday—I called on the president of the Belcher Lumber Company.

As it turns out, it was a very good idea of yours to have me go to Belfair. I only wish you had been more definite in your instructions in the first place, instead of merely requesting me to go in such a half-hearted manner. Fortunately, however, you braced up and got me down there before it was too late.

My interview with the president of the lumber company was very satisfactory, and ended with his having me ejected from the office. Before he did so, however, he lost his temper a bit, and blatted out a few remarks that he now probably wishes he had left unsaid.

The whole affair was most interesting and delightful. As soon as I had introduced myself to this lumber king, I started in on a description of the advantages of Earthworm tractors. He interrupted to say that on July twelfth he had been interviewed by Mr. Abner Smith, of Baxter Harbor, and had promised him that if he bought any tractors they would be Emperors. This statement, of course, only stimulated me to a greater flow of sales talk—which seemed to irritate the big lumber man more and more.

Finally, he got pretty sore. "Mr. Botts," he said, "I have heard all about you in a recent letter from Mr. Smith. Apparently you thought you could sell a lot of tractors to the state highway commission by working directly against me and in favor of that Baxter Forest park-and-highway project. And now, when the project looks as if it would fall through, you have the nerve to come around here and try to sell your tractors to me. I tell you, Mr. Botts, you can't play both ends of this thing and expect to get by with it. I don't care to discuss the matter anymore with you."

With these words he called in the janitor—a very large man—and had him escort me out of the building.

When I reached the sidewalk I was in a very thoughtful mood, turning over and over in my mind the somewhat cryptic remarks of the president of the Belcher Lumber Company. An ordinary man would probably have been completely baffled, but I have a considerable gift for spotting the ulterior motives of my fellow human beings, and before long I began to see a faint glimmer of light. It occurred to me that I ought to find out a little more about the business activities of this Belcher outfit.

Accordingly, I called on the cashier of what seemed to be the most important bank in town. I told him I was thinking of selling the lumber company a number of tractors, and I asked if he could tell me something of their credit rating and business standing. The cashier assured me that

"He called in the janitor—a very large man—
and had him escort me out of the building."

their credit was very high, and he very kindly gave me a copy of their annual report for stockholders, published about a month ago. Most of the stockholders reside in the city of Belfair. So it is probable that no one in Baxter Harbor—with the possible exception of Mr. Abner Smith—had seen or heard of this report.

As soon as I had read it, I decided that I had better be getting back here. I hurried down to the railroad station and caught a train which got me to Baxter Harbor last night just in time for the big public hearing on the highway project.

At the beginning of the meeting the chairman announced that all speakers would be limited to five minutes. This handed me a pretty heavy jolt, as the address which I had so carefully written out was designed to take up about an hour and three-quarters, not counting a long postscript which I had written on the train, and which discussed the Belcher Lumber Company.

For some time after the chairman had announced his outrageous gag rule, I was unable to collect my thoughts. All I could do was to sit there in more or less of a daze and contemplate with sorrowful eyes the voluminous manuscript which I had prepared with so much loving care. I heard very little of the first few speeches, which was probably just as well, as they were made by a lot of incredible nitwits—including Mr. Quackenbush and Mr. Abner Smith—and consisting almost entirely of ignorant and irrelevant remarks about hot-dog stands and billboards. Finally Mr. Dudley, the engineer, arose, and by that time I was able to concentrate sufficiently so that I could pay attention to his remarks.

Much to my surprise, he spoke very well—stating clearly and simply all the facts which he had given me in our interview some days ago. Most of the audience seemed very favorably impressed.

After Mr. Dudley finished speaking I decided it was up to me to do something. I arose, secured the recognition of the chairman, regretfully tossed my monumental manuscript out of a nearby window and stated that I would first read a very interesting paragraph from the annual report of the president of the Belcher Lumber Company, and that I would then explain to them exactly why Mr. Abner Smith had been so active in his opposition to the Baxter Forest highway-and-park project. The paragraph from the lumber company's report was as follows:

"Last year, when prices were at rock bottom, your directors quietly secured options on various tracts of timber land, amounting to approximately forty thousand acres, and comprising about half of what is known as the Baxter Forest, situated just north of the town of Baxter Harbor. There is now a proposal that the state of Maine condemn for park purposes the entire Baxter Forest, including the land on which we have options. A decision in this matter is expected within a few weeks. If the proposal goes through, our options will be void. If not, we will immediately take up this land and start lumbering operations. As the timber is mature and very high grade, we expect to utilize more than 90 percent of the trees. And as the purchase price will be very low, and as the price of the lumber is going up, we expect to realize a very handsome profit."

After I had finished reading this paragraph, there was an excited buzz of conversation. And, much to my surprise, Mr. Abner Smith leaped from his seat, rushed over to where I was standing, and began whispering excitedly into my left ear.

"What are you going to tell them about me?" he asked.

"It was my intention," I said, "to inform them that if the highway project fell through, so that the lumber company could go ahead and destroy the Baxter Forest, you had the promise of a very fat and profitable order for the sale of five big Emperor tractors."

"But you can't tell them that!" he said. "They would misunderstand. They would think I was not sincere in opposing this highway. They would think I did it just to make a profit."

"Well, you did, didn't you?" I asked.

"Yes, but it would never do to tell these people about it. They are my neighbors and friends. They would think I was a crook and a hypocrite. It would ruin my reputation forever. You wouldn't want to do that. Think of my wife and children."

"Well," I said, "what do you want me to do about it—just lay down and let this highway be defeated?"

"No," he replied. "If you'll just keep your mouth shut, I'll explain this thing in my own way, and I'll make a speech that will be better for your side than anything you could do."

"All right," I said with my characteristic magnanimity. "I am not a man who harbors grudges. I have no desire to ruin you. I will give you a chance to see what you can do. And your speech had better be good. If it is not, I will just naturally spill your beans all over town." Having said this, I sat down. And Mr. Abner Smith addressed the meeting.

The first part of his speech was pure fiction, but the latter part was fairly sensible. "In the matter of this parkway project," he said, "I have been actuated by the highest and most unselfish motives. My one desire has been to preserve the wild life and the natural beauty of this region. But I now realize that I was completely misinformed as to the facts. I did not know that this project would be free from all roadside nuisances. I did not know that the alternative to the proposed highway and park would be the complete destruction of our beautiful forest. But now that I have heard Mr. Dudley's clear explanation of his plans, and now that my very dear friend, Mr. Alexander Botts, has divulged the nefarious plans of the Belcher Lumber Company, I must reverse my position. I realize that the only way to save the forest is to make it a state park. And I recognize the fact that if all the people are taxed to create this park, it will be only fair to have a road so that all the people can use it; without a road, it would be limited to those who have the time and physical strength to walk over the trail.

"We nature lovers who use the trail must remember that there are other sincere nature lovers who cannot use it on account of age, infirmity or other reasons. We must not be so selfish as to deny the enjoyment of the wilderness to such people as the unfortunate one-legged man whose letter you have no doubt read in the paper. Furthermore, a road of the kind Mr. Dudley proposes would be no more of a desecration of the wilderness than the trail which is already there. Accordingly, Mr. Chairman, I move that this meeting pass a resolution in favor of the proposed Baxter Forest highway and park."

After Mr. Smith sat down, the resolution was promptly seconded by Mr. Quackenbush, the senile president of The Baxter Trail Club. Before a vote could be taken, there were several irate protests from a number of my friends in the filling station and hot-dog business, who were completely disgusted when they found out that they had been supporting a highway which would offer no opportunities for their respective businesses. But they were a hopeless minority, and the resolution was passed overwhelmingly.

That was last night. This morning the state-highway people at Augusta gave their final approval to the highway project. And this afternoon Mr. Dudley signed the enclosed order for ten seventy-five-horsepower diesel Earthworm tractors.

Yours,
ALEXANDER BOTTS.

THE BIG TREE

ILLUSTRATED BY TONY SARG

TELEGRAM
BRIDGEPORT CONN DEC 12 1933
GILBERT HENDERSON
SALES MANAGER
EARTHWORM TRACTOR CO
EARTHWORM CITY ILL

ARRIVED HERE YESTERDAY AND FOUND JAMES DODD
OUR LOCAL DEALER ALL EXCITED OVER VERY LIVE
PROSPECT WHO HAD PROMISED TO BUY TRACTOR IF WE
WOULD DEMONSTRATE BY MOVING BIG TREE FOR HIM
STOP PLANNED TO DO THIS TODAY BUT LAST NIGHT
SOME DIRTY BUM STOLE THE TREE STOP PLEASE WIRE
ONE THOUSAND DOLLARS TO BE OFFERED AS REWARD
FOR ITS RETURN STOP WE MUST ACT QUICK TO AVOID
SERIOUS EMBARRASSMENT

ALEXANDER BOTTS

TELEGRAM
EARTHWORM CITY ILL DEC 12 1933
ALEXANDER BOTTS
CARE JAMES DODD
EARTHWORM TRACTOR DEALER
BRIDGEFORT CONN

YOUR WIRE DOES NOT SOUND REASONABLE TO ME STOP
PLEASE SEND LETTER GIVING FULL DETAILS

GILBERT HENDERSON

ALEXANDER BOTTS
SALES PROMOTION REPRESENTATIVE
EARTHWORM TRACTOR COMPANY

CARE JAMES DODD,
EARTHWORM TRACTOR DEALER,
BRIDGEPORT, CONNECTICUT.
TUESDAY EVENING, DECEMBER 12, 1933.

MR. GILBERT HENDERSON,
SALES MANAGER,
EARTHWORM TRACTOR COMPANY,
EARTHWORM CITY, ILLINOIS.

DEAR HENDERSON: I am much disappointed in your wire. The situation here is very complex, and the thousand dollars is needed at once. I could not give you all the reasons in a telegram, but I hoped you would trust my judgment. I am right here, in close touch with the whole affair, I have all the facts in front of me, and I know what ought to be done. On the other hand, you are cooped up in a far-away office, completely out of touch with the world of reality. You have nothing in front of you but a row of silly filing cases, and it is clearly impossible for you to make a wise decision as to what ought to be done.

However, as you insist on trying to decide things yourself, I will give you a complete account of this affair—sending it to you by air mail, special delivery, in the hope that you may wire me the money before it is too late.

I arrived in Bridgeport in the middle of a snowstorm yesterday afternoon—Monday, December eleventh—and at once called on our local dealer, Mr. James Dodd. After introducing myself, I gave him the usual line to the effect that I had come to spend a few days with him, advising him regarding the latest sales methods, and helping him to reenergize and speed up his business. He reported that sales were slow, but better than last year, and he said that at the moment he had a very live prospect in the shape of an elderly gentleman called Mr. Algernon Cooper. It seems that this guy has a big shore-front estate about ten miles west of here, near a town called Westport. He has been considering buying one of our largest tractors to use next spring for some extensive grading around the place. And he told Mr. Dodd that he would buy the machine right away, instead of waiting for spring, if Mr. Dodd would take it over there and move a big tree for him.

Mr. Dodd naturally agreed to this, and he has been waiting for more than a week—ever since Mr. Cooper made him the proposition—until the weather would be cold and snowy enough. The weather is important on account of the peculiar technic which was to be used in this heavy transplanting operation.

It appears that the tree which Mr. Cooper wanted to move was a huge and truly magnificent specimen located on a small point of land jutting out into Long Island Sound at the extreme southwestern end of Mr. Cooper's big estate. In recent years the waves have been wearing away this point and threatening the very existence of the tree. And it seems that this situation has been very disturbing to old Mr. Cooper because, for various sentimental reasons connected with a romance of his early youth, he has a very deep affection for this particular giant of the woods. At first he thought of building a sea wall to protect it, but finally he decided to move it something over half a mile to his palatial residence at the opposite end of his estate. By planting it directly in front of his window, he will have it where he can see it first thing every morning when he wakes up.

The preliminary work for this monumental transplanting job was done last fall. A deep trench was dug around the tree, and, as soon as cold weather came and the earth began to freeze, the digging was gradually extended under the roots until they were completely undermined. The tree was then all ready to be moved, with its roots safely encased in a gigantic ball of solidly frozen earth. As an added precaution, this ball of earth was tightly bound around with heavy burlap and rope. One side of the hole was dug away to form a gentle incline up which the tree could be skidded. And a huge sledge, something like an overgrown stone boat, was especially constructed to carry this enormous burden on the half-mile trip to the house.

All these preparations had been completed several weeks ago, and the only thing which held up the actual moving was the fact that there was no snow to provide a proper surface over which the sledge should be dragged. On the afternoon of my arrival in Bridgeport, however, a very brisk little snowstorm was underway.

During the course of my conference with Mr. Dodd, the snowflakes descended in ever-increasing volume, until the ground was covered to a depth of two or three inches, and the storm showed no signs of letting up.

"It looks as if we could pull off the big job tomorrow," said Mr. Dodd. "You had better come out and see the show."

"I will be there," I said.

Mr. Dodd then showed me a large file containing information about various prospective purchasers, and suggested that I look it over. It was by this time five o'clock, and Mr. Dodd wanted to go home, so he showed me how to put out the lights and lock up, and then he departed, leaving me alone in the office.

I spent about two hours looking over the file, without finding anything of much interest, and a little after seven I closed it up and decided to go over to the hotel for supper. Just as I was starting to turn out the lights, however, there came a loud knocking at the front of the office. I opened the door end admitted a large, rough-looking man with a somewhat peculiar face. He brushed off the snowflakes from his big overcoat and handed me a carburetor with a badly cracked bowl.

"This carburetor," he said, "is off one of your Earthworm tractors. I want to buy a new one just like it, only without the crack."

"Very good, sir," I said. "I will see if I can find one for you."

I went back into the parts department and pawed around among the shelves and bins for about fifteen minutes, but I had no luck. I called up Mr. Dodd, but there was no answer—evidently, he was out for the

"I opened the door end admitted a large, rough-looking man with a somewhat peculiar face."

evening. When I told my visitor that I could not supply him with a new carburetor, he seemed much disturbed.

"I just started an important moving job out in the country," he said, "and one of the boys let a timber fall against the tractor and it cracked the carburetor. So we are all tied up. And we have got to get started again right away."

"Tonight?" I asked.

"Tonight," he replied.

"All right," I said. "I am a traveling representative of the Earthworm Tractor Company; I have general orders to help all customers in distress. I'll repair this broken carburetor the best I can. and I'll go out on this job with you and make it work."

The guy hesitated slightly. "I am not so sure," he said, "that I want you sticking your nose into this affair."

"Why not?"

"Well," he said, "I always like to handle a job like this by myself. But wait a minute. Did you say you are a traveling representative of the tractor company?"

"Yes."

"Are you acquainted around here at all?"

"No."

"In that case," he said, "I guess it would be safe to take you. Let's get going."

I did a quick repair job on the carburetor by winding the cracked bowl with some friction tape which I found in the parts room. Then I closed up the office, and my impatient customer drove me away in a large sedan. Before long we were outside the city and proceeding along a broad main highway. In spite of the snow, which was still falling fast, I was able to make out a few signs which indicated that we were on the Boston Post Road, headed for Westport, Norwalk, Stamford, and New York.

"I believe you said this is a moving job," I remarked. "Is it a house?"

"No," said my companion.

As he did not seem very communicative, I politely refrained from asking him anymore questions. But after long thought he finally reopened the subject.

"I suppose I might as well explain this thing to you," he said. "I am moving a large tree."

"But why do you have to do it at night?"

"I thought you would ask me that," he said. "The answer is very simple. I have to move this tree along the main highway for some distance, and as the tree is very large, it will block traffic to some extent. In order to get a permit for the job, I had to agree to do it at night. That sounds reasonable enough, doesn't it?"

"It certainly does," I agreed.

We rode on. After about half an hour, we turned off the main road into what seemed to be a private road. We followed this road for perhaps a quarter of a mile, and then stopped.

"Here we are," said my companion.

I at once alighted from the car, and found myself confronted by a most unusual scene. Directly ahead of me was a very tall and stately tree, brightly illuminated by a powerful electric spotlight mounted on an Earthworm tractor which was standing nearby. The tree seemed to be growing out of a huge ball of earth all wrapped about with rope and burlap, and resting in the bottom of a big hole in the ground. One side of the hole was dug away so as to make a gentle incline leading up to the place where the tractor stood. One end of a heavy cable was fastened around the roots and the other end was attached to the drawbar of the tractor. Additional cables were fastened to the branches of the tree, in such a way that they could be used for steadying the tree when it was moved. There were several workmen lurking in the shadows. I also noted that a large sledge, built like a stone boat, had been placed ready to receive the tree at the top of the incline loading up out of the hole.

I walked over to the tractor, attached the carburetor to the intake manifold and connected up the fuel lines. I then turned on the gasoline, and was delighted to find that the somewhat crudely taped bowl did not leak a single drop.

When we cranked the old baby up, however, it seemed to be very difficult to get her started. After considerable priming and cranking, the motor finally got going, but it ran very irregularly—sputtering and coughing back through the carburetor, and finally going dead on us. I opened the needle valve and tried again, using the choke very freely, but it was no use. Evidently, the interior works of the carburetor—the float, or the float-valve mechanism—had been knocked cockeyed at the time the bowl was cracked, and she just wasn't getting the gas.

I began to feel that I was in a fairly tough spot. There I was, shrouded in the darkness of night, with no adequate protection from a driving snowstorm, with my fingers stiff from the cold, and far from the facilities of any repair shop. It was obvious that this was neither the time nor the place to disassemble a carburetor and attempt to repair its delicate working parts. It was a situation, in short, which would have defeated any average man. But Alexander Botts has not spent fifteen years in the tractor business for nothing.

"If we can't feed her the gas in one way," I said, "we will feed it to her in another."

One of the workmen, at my request, dug a long-nosed squirt can out of the tool box. I poured out the lubricating oil which it contained, and filled it with gasoline from the tank on the tractor. I then climbed up onto the side of the machine and yelled to one of the workmen to crank the motor. When it started up, I began squirting gasoline into the air intake of the carburetor, and everything worked fine. I listened to the motor with the greatest intentness. At the first sign of heaviness in the explosions, I would squirt slightly less vigorously. At the first sign of coughing or popping back, I would squirt slightly more voluminously. Before long I found I was regulating the flow with the greatest precision, and the motor was running almost as smoothly as if it had had a perfectly adjusted carburetor.

As soon as I had shouted out the glad news that all was well, the man who operated the tractor leaped into the seat, and the big tree-moving job got underway. Working very slowly and carefully, with many stops while the various cables, props and timbers were readjusted, we began to drag that mighty tree up out of the hole.

One of the men discovered another oil can, which he oiled with gasoline to use as a spare. As soon as one can became empty he would hand me up a full one to replace it. In this way I was able to keep up a continuous squirting, and there was no delay on account of refilling cans. But the job was a long one, nevertheless. The slightest carelessness would have resulted in tipping over the big tree, so we did not dare hurry, and it was three or four hours before we had the old baby safely skidded onto the big stone boat at the top of the incline.

During all this time I had been clinging to the side of the machine in much the same way as a monkey clings to a coconut tree. As a result, I was almost exhausted—which goes to show that a highly intelligent human being, such as myself, has very poorly developed tree-gripping muscles and is sure to get into trouble whenever he tries to imitate a monkey, ape or other anthropoid. In addition to being all tired out, I was hungry—having had no supper. And I was practically frozen to death, besides, my overcoat was none too heavy, my feet were protected only by medium-weight oxfords, and the chilliness of the weather was more than sufficient to overcome the boat generated by the motor.

As soon as the tree had been safely lashed onto the stone boat, I climbed down from my uncomfortable perch and had one of the workmen take my place with the squirt can. Fortunately, he turned out to be a bright lad, and by following my instructions he soon got the hang of the thing very well. Now that we were upon level ground, we were able to go very much faster, and in less than an hour we had almost reached the point where the private road entered the main highway. At this point, the big boss—the man who had brought me out from Bridgeport—came up to me as I was feebly and wearily dancing about, so as to keep my feet warm, and pitifully attempting to prevent my hands from freezing by slapping them against my shoulders.

"I think," he said, "that from now on we can get along without you. If you will get into my car, I will take you back to Bridgeport.

"I hate to seem like a quitter," I said.

"You've done a swell job out here. If it hadn't been for you, we never could have moved this tree. We couldn't even have got away with the tractor. You don't realize it now, but later on you will come to understand how important your help has been to me. You have my most heartfelt thanks."

"I had been clinging to the side of the machine in much the same way as a monkey clings to a coconut tree."

"From now on we can get along without you."

"Thank you," I said. "But are you sure you will be able to get along without me?"

"Absolutely," he replied. "You have shown us how to work this thing, so now we can handle it ourselves."

"Well," I said, "as long as you feel sure you can go ahead without me, and as long as I seem to feel the ice crystals beginning to form in my blood stream, perhaps I had better take your advice."

We climbed into the car, which one of the men had brought along in the wake of the tractor and the tree, and a half hour later I was deposited at the hotel in Bridgeport. It was then a little past midnight.

"How much do I owe you?" asked my companion.

"Nothing at all," I said. "This is just a sample of the service given by the Earthworm Tractor Company."

The guy thanked me once more, and drove off.

After taking a hot bath, and after further warming myself up with some refreshments which the bell boy procured for me, I tumbled into bed, when my weary frame lapsed into such a deep slumber that I did not wake up until noon.

I arose, feeling very much refreshed, and after fortifying myself with a double order of ham and eggs, coffee and other things, I started down to

Mr. Dodd's office. The storm was now over. The sky was blue, and the snow sparkled prettily in the bright winter sunshine. The brightness was not all external either; I was aware of a considerable amount of sunshine in my soul as I recalled the happy and successful outcome of my arduous labors. And, considering the difficulties I had encountered, you will have to admit that I had done a swell job, and had ample reason to be very proud of myself. As I entered Mr. Dodd's office, I was looking forward most pleasantly to the words of praise which I supposed would fall from his lips when I told him what good deeds I had accomplished.

Strange as it may seem, however, I never got to tell Mr. Dodd about these good deeds, because, as soon as I entered the office, I was given some information that made me decide I had better keep my mouth shut. I found Mr. Dodd in the midst of a very excited conversation with an old gentleman whom he introduced to me as Mr. Algernon Cooper.

"Mr. Cooper has just told me," he said, "that somebody sneaked into his place last night and stole the big tree which we were planning to move for him today."

"It sounds incredible," I replied.

"It's true," said Mr. Cooper, "and it happened last night. That tree was there yesterday; I saw it with my own eyes. But when I walked down there this morning to see whether the snow was deep enough and whether everything was ready for the moving, it was gone. The stone boat was missing too. I think somebody must have come along with a big tractor and hauled it away."

"Did you find any tracks?" I asked.

"No," said Mr. Cooper. "It snowed practically all night, so any tracks that may have been there are completely obliterated."

"Then how do you know it was a tractor?" I asked. "Nobody saw these people take the tree, did they?" I asked this last question with a feeling of considerable anxiety.

"As far as I know," said Mr. Cooper, "nobody saw anything. But for several hours last night I kept hearing the sound of a powerful motor somewhere. I paid very little attention to it at the time. I thought that it might be a snow plow working on the Post Road. But now I am sure it must have been thieves taking away my property."

"But I never heard of such a thing as stealing a great big tree," I said.

"It has been done," said Mr. Dodd. "These wealthy estate owners around here will often pay as much as several thousand dollars apiece for full-sized shade trees. And, sometimes, when the proper kind of tree is hard to get, there have been crooked nurserymen who have stolen what they needed. I believe this is exactly what has happened in this case. What do you think about it, Mr. Botts?"

Now, I might as well admit to you personally, Henderson, that by this time I had begun to have some very definite theories on this matter. But the more I considered it, the more I became convinced that the part of wisdom would be for me to conceal my suspicions for the time being.

"This is indeed a baffling mystery," I said, in a quiet and thoughtful tone of voice, "You don't suppose your tree could have been merely blown down and covered up with snow, do you."

"It would take a drift forty feet deep," said Mr. Cooper.

"Let me see," I said. "It was standing very near the shore, wasn't it? Possibly it fell over into the water and floated away."

"How could it float away," said Mr. Cooper, "when its roots were frozen solid into a cake of earth as big as a small house? No," he continued, "it was stolen. I came over here on the off chance that you people might have decided to start the moving job ahead of time without notifying me. But now, as long as you don't know anything about it, the only thing to do is to call in the police."

"Oh, I wouldn't do that," I said.

"Why not?"

"Well," I said, "if you notify the police, everybody will hear about it. And I wouldn't think a man of your position would want to be involved in a public scandal."

"How will it involve me in any scandal to have it known that I have been robbed by a bunch of crooks? Hand me the telephone."

He called the police station and explained the whole situation. Then he turned back to me.

"I am going to meet the police officers out at my place right away," he said. "These criminals ought to be easy to trace. They can't move a big tree like that very far without being seen, so we ought to capture them pretty quick. And when we do, there will be plenty of scandal—just as you suggested—but the only person involved in it will be the crook that stole the tree, and the other crooks, if any, that helped him."

With these words Mr. Cooper took his departure. And, while Mr. Dodd went to the door to wish him goodbye, I remained in the office to ponder over the disagreeable thoughts aroused by Mr. Cooper's last remark regarding the scandal which would attach to anyone who had assisted in this unfortunate robbery.

It seemed probable, with a bunch of cops hot on the trail, that my lawless customer of the night before would soon be caught. And as soon as he realized that they had the goods on him, he would probably tell all—in which case the Earthworm Tractor Company would be revealed as an accomplice.

Having reasoned things out to this point, I immediately went down to the telegraph office and sent you my wire requesting a thousand dollars—which request you refused. Now that you know the facts in the case, however, I am sure that you will grant my wishes. What I want to do is to put an advertisement in the paper addressed to this nefarious tree snatcher. I will point out that the police are right after him, that he cannot hope to elude them much longer, but that we are ready to give him one last chance. If he will bring back the tree, we will pay him a reward of one thousand dollars, and no questions asked.

I have a very strong hunch that this advertisement will bring back the stolen property. And, if it does, Mr. Cooper will be completely satisfied. All he wants is to recover his tree, for which, as I explained before, he has a great sentimental attachment. As soon as he recovers it, he will undoubtedly consent to call off the cops. This will remove all danger of involving the company in scandal, and it will, therefore, be well worth the thousand dollars. I trust that you will wire this amount without delay. In my opinion, it is the only sure way of handling this unfortunate situation.

While waiting for the money, however, I have been working along as best I can without it. As soon as I had sent the telegram, early this afternoon, I hired a small car and started a private investigation. I first drove out to Mr. Cooper's place, where I checked over the ground and proved beyond the last shadow of a doubt that it was indeed his tree which I had helped carry off. I then spent the rest of the afternoon driving along the Post Road and the neighboring side roads, asking all the inhabitants whether they had seen a large tree going by. All the answers were in the negative. But tomorrow I am going out again and I will keep right on, covering an ever-widening area, until I find something. If I can only catch up with these tree snatchers before the police get them, I may be able, by means of threats and persuasion, to get them to take back the tree secretly, and thus avoid the public scandal which I fear.

"Up to this time, the police have apparently had no better luck than I have. The snowstorm last night seems to have provided a curtain of falling flakes so dense that no one observed the movements of the criminals. And these same flakes, of course, covered up their tracks as soon as they were made. That tree may be as much as fifteen or twenty miles away by this time. But, even so, it is sure to be found in the end. And it is most essential that we find it before the police.

And this brings me right back to my principal argument. The only sure way to reach these people ahead of the cops is by offering that thousand dollars reward, and, in closing this letter, I want to urge you once more to wire me the money as promptly as possible. You don't want to deceive yourself about where the blame will be placed, in case all the facts come out. It is true that I did the actual work of assisting in this robbery, but no blame can attach to me personally. I was acting under general orders from the company in assisting an Earthworm tractor operator. And this fact quite clearly throws the entire responsibility on the company which issued those orders.

Most sincerely,
ALEXANDER BOTTS.

P.S. If you write Mr. Dodd, don't tell him I helped move away that tree.

———

NIGHT LETTER
EARTHWORM CITY ILL DEC 15 1933
ALEXANDER BOTTS
CARE JAMES DODD
EARTHWORM TRACTORS DEALER
BRIDGEPORT CONN

YOUR LETTER RECEIVED BUT YOU HAVE NOT YET GIVEN FULL DETAILS STOP WHAT KIND OF TREE WAS IT QUESTION MARK HOW BIG QUESTION MARK WHAT DID IT LOOK LIKE QUESTION MARK WHY DOES OWNER HAVE SENTIMENTAL ATTACHMENT FOR IT QUESTION MARK IS MR DODDS SUPPOSEDLY COMPLETE STOCK OF PARTS ACTUALLY SHORT ON CARBURETORS QUESTION MARK IF SO WHY QUESTION MARK WHEN YOU SAY THAT THE CHIEF TREE SNATCHER HAD A PECULIAR LOOKING FACE EXACTLY WHAT DO YOU MEAN QUESTION MARK DID YOU TAKE DOWN THE MOTOR NUMBER OF THE

TRACTOR USED IN THE ROBBERY QUESTION MARK IF NOT WHY NOT QUESTION MARK.

GILBERT HENDERSON

DAY LETTER
BRIDGEPORT CONN DEC 15 1933
GILBERT HENDERSON
EARTHWORM TRACTOR COMPANY
EARTHWORM CITY ILL

CANNOT UNDERSTAND WHY YOU INSIST ON QUIBBLING ABOUT UNIMPORTANT DETAILS IN FACE OF GREAT EMERGENCY STOP THE TREE WAS A VERY FINE SPECIMEN OF FAGUS SYLVATICA COMMONLY KNOWN AS BEECH STOP VERY TALL BUT I DID NOT MEASURE IT STOP OWNER VERY SLOPPY AND SENTIMENTAL ABOUT IT BECAUSE MANY YEARS AGO HE CARVED ON IT A LARGE HEART WITH HIS INITIALS A C AND THOSE OF HIS GIRL N K STOP THIS CARVING STILL DISTINCT ALTHOUGH HE AND THE GIRL HAVE BEEN MARRIED MANY YEARS STOP MR DODD HAD NO CARBURETOR TO FIT THE MACHINE USED IN ROBBERY AS IT WAS AN OBSOLETE SIXTY HORSEPOWER NINETEEN TWENTY FIVE MODEL AND HE DID NOT KNOW THAT THERE WERE ANY OF THESE MACHINES LEFT IN HIS TERRITORY STOP CHIEF ROBBER HAD VERY LARGE NOSE BENT SLIGHTLY SIDEWISE AS FROM FORMER ACCIDENT STOP I DID NOT TAKE DOWN MOTOR NUMBER BECAUSE TOO DAM BUSY AND TOO DAM DARK STOP HOPE THIS SATISFIES YOU STOP I AGAIN WISH TO REMIND YOU MOST RESPECTFULLY THAT THE MAN ON THE JOB WHO IS IN ACTUAL CONTACT WITH THE FACTS KNOWS MORE THAN ANYBODY IN AN OFFICE WHO IS IN CONTACT ONLY WITH A LOT OF DUSTY MUSTY FILES STOP IT IS ESSENTIAL FOR THE GOOD OF THE COMPANY THAT YOU TRUST MY JUDGMENT AND WIRE THAT THOUSAND DOLLARS AT ONCE

ALEXANDER BOTTS

TELEGRAM
EARTHWORM CITY ILL DEC 15 1933
ALEXANDER BOTTS
CARE JAMES DODD
EARTHWORM TRACTOR DEALER
BRIDGEPORT CONN

HAVE DECIDED THAT THE EARTHWORM TRACTOR
COMPANY IS UNDER NO OBLIGATION TO SPEND ANY
MONEY ON THIS TREE ROBBERY STOP RESPONSIBILITY
RESTS ON YOU PERSONALLY STOP BUT AS LONG AS
YOU WERE INNOCENT OF ANY EVIL INTENT I HOPE
EVERYTHING MAY WORK OUT ALL RIGHT IN THE END

GILBERT HENDERSON

ALEXANDER BOTTS
SALES PROMOTION REPRESENTATIVE
EARTHWORM TRACTOR COMPANY

CARE JAMES DODD,
EARTHWORM TRACTOR DEALER,
BRIDGEPORT, CONNECTICUT.
TUESDAY, DECEMBER 10, 1933.

MR. GILBERT HENDERSON,
SALES MANAGER,
EARTHWORM TRACTOR COMPANY,
EARTHWORM CITY, ILLINOIS.

DEAR HENDERSON: I have good news for you. Single-handed and absolutely alone, without any help from the police or anyone else, and in spite of the complete and well-nigh incredible lack of all cooperation from the Earthworm Tractor Company, I have accomplished the astounding feat of bringing back the stolen tree. And the whole affair was handled so smoothly and cleverly that everybody is satisfied, and not even the faintest breath of scandal has touched the Earthworm Tractor Company or myself.

As soon as I received your disappointing telegram on Saturday, I realized that everything depended upon me. So I inserted an advertisement in the Sunday-morning paper offering the tree snatchers fifty dollars out of my own pocket if they would bring back the stolen property, and threatening them with arrest if they did not. The fifty-dollar reward was, of course, absurdly inadequate as compared to the thousand dollars I had hoped to offer. And the threat of arrest was a good deal of a bluff, as neither the police nor I had been able, up to that time, to uncover any clues that amounted to anything. But I was doing the best I could.

After arranging for the advertisement, I continued my investigation in the little rented car: driving over all the main and secondary highways in the territory adjoining the scene of the crime, and questioning literally thousands of the inhabitants.

About eleven o'clock last night Mr. Cooper received a mysterious telephone message telling him that he would find his tree on the road to Johnson's Landing. This road is a short one-track affair which starts from an abandoned wharf on the shore of the Sound, and joins the Post Road at a point several hundred yards beyond the road over which the big tree had been dragged on the night of the robbery.

Mr. Cooper, being a vigorous old gentleman, at once rushed out and made his way to the place indicated. And there, resting solidly on the big sledge, stood his tree—beautiful as ever and completely unharmed. It had evidently been put there only a short time before, as both the police and I had gone over the Johnson's Landing road several times without finding even a single clue.

This morning Mr. Cooper conveyed the glad news to Mr. Dodd, and tomorrow we are going to move the tree up to Mr. Cooper's house.

The old gentleman has notified the police that he is no longer interested in pursuing the criminals, so it is probable that they will never be known. There can be no doubt, however, as to why they returned the stolen property. The only possible explanation is that at some time during my long and tortuous journey of investigation I must have approached, without realizing it, so close to my quarry that they began to fear the game was up. This fact, coupled with the threatening remarks in my advertisement, must have scared them so thoroughly that they sneaked back with the stolen tree, and then disappeared without even having the nerve to ask for the reward—which is very pleasing, as it saves me fifty dollars.

Of course, it is true that there will always remain down through the ages a deep and impenetrable curtain of mystery which will forever keep us from knowing the answers to such questions as: What is the name of the man who stole the tree? Where did he take it? How did he keep it concealed for so many days? What had he planned to do with it? And how did he get it back over the busy roads of Southern Connecticut without being detected? But, although all these questions will never be answered, it is perfectly clear that the credit for recovering the stolen property belongs entirely to your proud and happy sales representative.

ALEXANDER BOTTS.

EARTHWORM TRACTOR COMPANY
EARTHWORM CITY, ILLINOIS
OFFICE OF THE SALES MANAGER

FRIDAY, DECEMBER 22, 1933.

MR. ALEXANDER BOTTS,
CARE JAMES DODD,
EARTHWORM TRACTOR DEALER,
BRIDGEPORT, CONNECTICUT.

DEAR BOTTS: Congratulations on the happy outcome of the affair of the stolen tree.

I have been much interested in this case and have tried to help you as much as I could. In your first account of the robbery, however, you omitted so many essential details that I was almost as puzzled and helpless as you were. It was for this reason that I asked for additional facts. You were kind enough to furnish them in your telegram of last Saturday. And, as soon as I knew that the criminal had a large crooked nose, that the tractor was a 1925 sixty-horsepower model and that the tree was a beech with certain definite carvings on it, I was ready to do something.

It occurred to me that your crooked-nosed friend would probably order a new carburetor to replace the one which you had partially patched up. With this thought in mind, I began pawing through the musty, dusty files, which—as you have pointed out—are my only contact with the world of reality. I soon discovered, from the records of the parts department, that practically all of our dealers have ceased carrying in stock the carburetor for the now obsolete 1926 sixty-horsepower model. There are so few of these machines working at the present time that sales of parts for them are very few and far between and are handled almost entirely out of the main office here.

Further inspection of the files revealed that only one carburetor for the 1925 sixty-horsepower model has been sold during the past six months. This one sale was made on a telegraphic order from our Long Island dealer, with headquarters at Brooklyn, New York, on Tuesday, December twelfth—the day after the robbery. The order requested that shipment be rushed, as the customer for whom it was intended was in a hurry.

I at once wired the Brooklyn dealer, giving him a description of the tree, the tractor and the crooked-nosed man, and asking him to investigate the matter.

In reply to this wire, the dealer has written me that late on Saturday afternoon he went out to see the man who ordered the carburetor. This man is a general contractor of rather shady reputation, and he has a large crooked nose.

The dealer found him at the entrance of a large estate on the north shore of Long Island. He was engaged in moving, by means of his 1925 sixty-horsepower tractor, a large beech tree upon whose smooth trunk was carved a heart with the initials AC and NK. The tractor was working smoothly with a new carburetor, and in the toolbox there was an old carburetor, wound around with friction tape.

When the contractor was confronted with all the evidence against him, he broke down and confessed everything. It appears that the owner of the estate which he was entering had promised him five thousand dollars for a beech tree which would meet certain specifications as to size and symmetry. The contractor had been unable to find any such tree on the north shore of Long Island. His brother, who lives in Connecticut, had told him of Mr. Cooper's tree—which was just what he wanted, and was all ready to move, but was, unfortunately, not for sale.

The unscrupulous contractor had thereupon loaded his car, his tractor, a lot of tools and a gang of workmen onto one of the large barges which he normally used for transporting sand, gravel and building supplies. He had used one of his small tugboats to tow this barge across the Sound to the abandoned wharf at Johnson's Landing, had moved the tree on board and had brought it back to Long Island.

As our Long Island dealer is extremely kind-hearted, and as he was dealing with a man who is a customer of his, he finally agreed not to report the matter to the authorities. In return, the contractor agreed to take the tree back as far as Johnson's Landing and notify the owner of its arrival. It seems that he carried out this agreement on Monday night. And, in addition, he has settled up a rather large and long-overdue account with our Long Island dealer—which, naturally, pleases the dealer very much.

I trust that the above information may be of interest to you.

Most sincerely,
GILBERT HENDERSON,
Sales Manager.

TELEGRAM
BRIDGEPORT CONN DEC 24 1933
GILBERT HENDERSON
EARTHWORM TRACTOR COMPANY
EARTHWORM CITY ILL

WE HAVE MOVED THE TREE AND MR COOPER HAS BOUGHT THE TRACTOR STOP I AM MAILING YOU CHECK FOR FIFTY DOLLARS WHICH I OFFERED ANYONE WHO WOULD BRING ABOUT THE RETURN OF THAT TREE STOP YOU CERTAINLY EARNED THIS REWARD STOP BEST WISHES FOR A MERRY CHRISTMAS AND A HAPPY NEW YEAR FROM SOMEONE WHO IS RIGHT ON THE JOB AND SEES EVERYTHING BUT KNOWS NOTHING TO AN INTELLECTUAL GIANT WHO HAS NOTHING TO WORK WITH BUT DUSTY MUSTY FILES AND WHO ACCOMPLISHED WONDERS JUST THE SAME.

ALEXANDER BOTTS

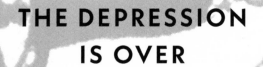

THE DEPRESSION
IS OVER

ILLUSTRATED BY TONY SARG

EARTHWORM TRACTOR COMPANY
EARTHWORM CITY, ILLINOIS
OFFICE OF THE SALES MANAGER

THURSDAY, JULY 12, 1934.

MR. OLAF ANDERSEN,
MANAGER EARTHWORM TRACTOR EXHIBIT,
CENTURY OF PROGRESS EXPOSITION,
CHICAGO, ILLINOIS.

DEAR MR. ANDERSEN: We are shipping you today by express one special high-speed rotary water pump designed to be attached to the rear end of the transmission case of our new one-hundred-horsepower Earthworm tractor.

This pump is a new model and will not be in regular production until the middle of next month. In order that you may have a sample of the pump to exhibit at the fair, we have assembled one out of such parts as are at present available. It is complete in every way except for the two steel gears in the case behind the pump chamber. As production on these gears has been delayed, we have had our pattern shop turn out two mahogany gears which have been installed in place of the steel ones. The mahogany gears have been coated with aluminum paint, so they look just like steel. And they are so accurately made that they operate with complete smoothness.

If you will remove the cover plates from the pump and gear case, you can give prospective customers a very effective demonstration by turning the gears by hand. Be sure that nobody tries to use the pump. I have attached a tag reading: "For exhibition purposes only. Not to be sold or attached to any tractor," and I would suggest that you leave this tag in place. But you had better not tell anyone that the gears are fakes. We do not want to advertise the fact that we are behind in our production. If anybody wants to buy a pump, you will order it through the factory, and shipment will be made as soon as possible.

I am enclosing a letter for Mr. Alexander Botts, our traveling representative. Mr. Botts has been in the East, and expects to arrive in Chicago and call on you within the next few days.

Very truly,
GILBERT HENDERSON,
Sales Manager.

EARTHWORM TRACTOR COMPANY
EARTHWORM CITY, ILLINOIS
OFFICE OF THE SALES MANAGER

THURSDAY, JULY 12, 1934.

MR. ALEXANDER BOTTS,
CARE MR. OLAF ANDERSEN,
EARTHWORM TRACTOR EXHIBIT,
CENTURY OF PROGRESS EXPOSITION,
CHICAGO, ILLINOIS.

DEAR BOTTS: Our business has recently been increasing to such an extent that we are planning a reorganization in the sales department, as well as in certain other parts of the company. As this may affect your position, we want you to come down to Earthworm City at once, so that we may talk matters over with you.

Very sincerely,
GILBERT HENDERSON,
Sales Manager.

———

ALEXANDER BOTTS
SALES PROMOTION REPRESENTATIVE
EARTHWORM TRACTOR COMPANY

CHICAGO, ILLINOIS.
MONDAY, JULY 16, 1934.

MR. GILBERT HENDERSON,
SALES MANAGER,
EARTHWORM TRACTOR COMPANY,
EARTHWORM CITY, ILLINOIS.

DEAR HENDERSON: I arrived in Chicago this morning. I am certainly glad to hear, from your letter, that the business of the Earthworm Company is increasing. It begins to look as if the depression is very definitely a thing of the past. Here at the World's Fair, everything seems to be booming.

The whole place swarms with people, and most of them look as if they had money and wanted to spend it. Our Earthworm tractor exhibit is crowded all the time, and a whole lot of the visitors are big contractors who are real sure-enough hot prospects.

As a matter of fact, business is so active that I will not be able, at the present time, to come down to Earthworm City as suggested in your letter. I am leaving for Cold River, Wisconsin, instead, for the purpose of selling one or more of our machines to a big contractor by the name of Duffield Watt. Mr. Watt stopped in at our exhibit at the fair soon after I arrived this morning. I discovered he was interested in buying some extra machinery for a large sewer-ditch-digging job which he is starting at the town of Cold River, Wisconsin. He had talked with our dealer at Cold River. And this dealer had apparently been so ineffective in his sales arguments that Mr. Watt had failed to buy, and had come down here to the fair in order to look over our exhibit and also the exhibits of all our competitors.

As soon as Mr. Watt told me this, I decided that I would have to get busy. If the Earthworm dealer at Cold River is such a sap that he weakly permits a live prospect to get out of his clutches and go wandering about the country investigating competing machines, it is high time for some-body like myself to step in and take charge of affairs.

Mr. Watt is returning to Cold River tomorrow morning. I am going along with him, and I will camp right on his trail until he buys at least one Earthworm tractor. You may address me in care of the Earthworm dealer at Cold River, Wisconsin.

Very sincerely,
ALEXANDER BOTTS.

———

TELEGRAM
EARTHWORM CITY ILL JULY 17 1934
ALEXANDER BOTTS
CARE EARTHWORM TRACTOR DEALER
COLD RIVER WIS

COME TO EARTHWORM CITY AT ONCE AS REQUESTED IN MY LETTER STOP LET LOCAL DEALER HANDLE SALE TO DUFFIELD WATT
GILBERT HENDERSON

NIGHT LETTER
COLD RIVER WIS JULY 18 1934
GILBERT HENDERSON
EARTHWORM TRACTOR COMPANY
EARTHWORM CITY ILL

LOCAL DEALER IS A MERE SPINELESS OYSTER INCAPABLE
OF HANDLING THIS IMPORTANT DEAL STOP I HAVE
INVESTIGATED MR WATTS DITCH DIGGING JOB AND FIND
HE IS HAVING TROUBLE WITH WATER SEEPING INTO
DITCH AND NEEDS TO PURCHASE PUMPING EQUIPMENT
STOP I AM SURE I CAN SELL HIM ONE OF THE TRACTORS
IN DEALERS STOCK HERE IF IT IS EQUIPPED WITH PUMP
STOP AND IF THIS WORKS WELL HE MAY BUY SEVERAL
MORE OUTFITS STOP PLEASE SHIP EXPRESS DOUBLE
RUSH ONE SPECIAL HIGHSPEED ROTARY WATER PUMP
FOR ONE HUNDRED HORSEPOWER EARTHWORM
TRACTOR JUST LIKE THE ONE ON EXHIBITION AT FAIR
IN CHICAGO

ALEXANDER BOTTS

TELEGRAM

EARTHWORM CITY ILL JULY 19 1934
ALEXANDER BOTTS
CARE EARTHWORM TRACTOR DEALER
COLD RIVER WIS

WATER PUMP AS ORDERED IN YOUR WIRE IS NEW
MODEL NOT YET AVAILABLE STOP WILL SHIP AS SOON
AS POSSIBLE PROBABLY ABOUT THREE WEEKS STOP
IMPERATIVE THAT YOU COME TO EARTHWORM CITY AT
ONCE

GILBERT HENDERSON

TELEGRAM

COLD RIVER WIS JULY 19 1934
GILBERT HENDERSON
EARTHWORM TRACTOR COMPANY
EARTHWORM CITY ILL

CANT WAIT THREE WEEKS STOP YOU WILL HAVE TO SHIP
ME THE PUMP ON EXHIBIT AT CHICAGO STOP PLEASE
SEND BY EXPRESS RUSH

ALEXANDER BOTTS

TELEGRAM

EARTHWORM CITY ILL JULY 19 1934
ALEXANDER BOTTS
CARE EARTHWORM TRACTOR DEALER
COLD RIVER WIS

CANNOT SHIP PUMP FROM FAIR AS IT IS FOR EXHIBITION
PURPOSES ONLY AND NOT FOR SALE STOP COME TO
EARTHWORM CITY AT ONCE STOP YOU MUST BE HERE
IN TIME FOR SPECIAL MEETING OF OFFICERS AND
DIRECTORS AT TEN OCLOCK SATURDAY MORNING JULY
TWENTY FIRST

GILBERT HENDERSON

ALEXANDER BOTTS
SALES PROMOTION REPRESENTATIVE
EARTHWORM TRACTOR COMPANY

COLD RIVER, WISCONSIN.
SATURDAY EVENING, JULY 21, 1934.

MR. GILBERT HENDERSON,
SALES MANAGER,
EARTHWORM TRACTOR COMPANY,
EARTHWORM CITY, ILLINOIS.

DEAR HENDERSON: I have very good news for you.

Your wire refusing to ship the pump from Chicago arrived yesterday morning—Friday. As soon as I had read it, I realized that you had no real comprehension of the urgency of the situation up here, and I therefore decided that I would have to get busy and go ahead on my own initiative.

Accordingly, I took the first train to Chicago. I arrived there late yesterday afternoon and went at once to our exhibit at the fair, where I informed Mr. Olaf Andersen that I had come to take away the water pump. I had anticipated no great trouble in talking him out of this piece of machinery, but, unfortunately, he seems to be as dumb and stubborn an old fellow as I have ever run across. All my arguments made no impression on him.

He first of all told me that he could not let me take the pump, and then he showed me a silly tag which some boob had fastened onto it. The tag read as follows: "For exhibition purposes only. Not to be sold or attached to any tractor." When I explained to him that my need for this pump was far more important than any mere exhibit, and asked him if there was any

"I informed Mr. Olaf Andersen that I had come to take away the water pump.
I had anticipated no great trouble in talking him out of this piece of . . .

real reason why he could not let me have it, he merely became evasive and said that he did not care to discuss the matter. Somehow, I don't like that guy—which is natural enough, because his failure to cooperate caused me a most annoying delay.

As a matter of fact, I had to wait until five o'clock the next morning—today—at which time I went out to the fairgrounds and carted away the pump in a little pickup truck, which I had rented for this purpose. I had an awful time talking my way past the guards at the gate and the watchman in the building where our exhibit is located. In the end, however, I managed to get away with my loot long before Mr. Andersen arrived to open up the exhibit. As soon as I was safely outside the fairgrounds, I headed north, and arrived up here in Cold River toward the end of the morning.

I at once got in touch with Mr. Duffield Watt, and asked him to come around to the dealer's showroom here. By the time he arrived, I had already installed the pump on the rear end of one of the machines which was in stock here. As soon as Mr. Watt saw the complete outfit, he decided that it was exactly what he wanted. I offered to demonstrate it for him, but he said that he would take my word for it that it would perform as advertised. He promptly wrote out his check for the full purchase price of both the tractor and the pump, the dealer gave him a bill of sale and then one of his mechanics drove the machine off in the direction of the big ditch which he is digging.

Tomorrow morning I will drive my rented truck back to Chicago, where I will await your further instructions. It seems useless for me to go on to Earthworm City, as the meeting which you wired me you wanted me to attend took place at ten o'clock this morning. Besides, there are a lot of things at the fair that I want to see, so I will just stick around there the next few days.

. . .machinery, but, unfortunately, he seems to be as dumb and stubborn an old fellow as I have ever run across. All my arguments made no impression on him."

My success in putting over the sale of the tractor and pump to Mr. Duffield Watt naturally fills me with a warm glow of satisfaction. In the first place, I am always particularly happy when I sell a machine which I am so absolutely sure will give complete satisfaction to the purchaser. And, in the second place, it is a real indication of returning prosperity when a man makes up his mind to buy a machine without any long period of negotiation and demonstration, and then pays cash in full. This certainly reenforces my opinion that the depression is over and that all our troubles are at an end at last. And, incidentally, Mr. Watt has stated that if this pump-and-tractor combination works out all right—which, of course, it will—he will buy four more similar outfits.

There is one unfortunate aspect to the situation, however. It seems rather too bad that in order to put over this deal I should have been compelled to demean myself; first, by having to argue with that dumb Andersen down at Chicago, and, second, by being compelled to remove by stealth a pump which should have been promptly turned over to me when I first asked for it. For years I have been vainly hoping that sometime the Earthworm Tractor Company might be able to organize itself so that matters of this kind could be handled by somebody who actually understands conditions in the field, rather than by a sales office which is more or less cut off from all actual contact with the customers. I don't want to seem critical or anything like that, but I sometimes have a feeling that it would be a wonderful thing if the office of sales manager were held by a really experienced and practical salesman—such, for instance, as myself.

Most sincerely,
ALEXANDER BOTTS.

"I had an awful time talking my way past the guards at the gate and the watchman in the building where our exhibit is located. . .

EARTHWORM TRACTOR COMPANY
EARTHWORM CITY, ILLINOIS
OFFICE OF THE PRESIDENT

MONDAY, JULY 23, 1934.

MR. ALEXANDER BOTTS,
CARE MR. OLAF ANDERSEN,
EARTHWORM TRACTOR EXHIBIT,
CENTURY OF PROGRESS EXPOSITION,
CHICAGO, ILLINOIS.

MY DEAR MR. BOTTS: I have the honor to inform you that the board of directors has appointed you to the position of sales manager of the Earthworm Tractor Company in place of Mr. Gilbert Henderson. I trust that you will send us, at the earliest possible moment, your acceptance of this office.

Cordially yours,
JOHN MONTAGUE,
President Earthworm Tractor Co.

... In the end, however, I managed to get away with my loot."

ALEXANDER BOTTS
SALES PROMOTION REPRESENTATIVE
EARTHWORM TRACTOR COMPANY

CHICAGO, ILLINOIS.
TUESDAY, JULY 24, 1934.

MR. JOHN MONTAGUE,
PRESIDENT,
EARTHWORM TRACTOR COMPANY,
EARTHWORM CITY, ILLINOIS.

MY DEAR SIR: I have the honor to inform you that I most emphatically refuse your offer of the position of sales manager of the Earthworm Tractor Company. In refusing, I may as well admit perfectly frankly that I would like nothing better than to have the job of sales manager, and I am pretty sure that I would be the best sales manager you ever had. But if you will give the matter a moment's thought, you must understand that my sense of loyalty to my old boss makes it absolutely impossible for me to climb into this office, figuratively speaking, over his dead body.

When I first received your letter, I was at a loss to understand why you should be offering me this promotion so suddenly, and without having given me even a hint of your intentions beforehand. Having thought the matter over, however, I have reached the conclusion that you and the directors of the company must in some way have got hold of the letter which I wrote to Mr. Henderson last Saturday. This letter contained such a clear statement of Mr. Henderson's shortcomings—as exemplified in his failure to send me that pump—and it contained such a convincing argument in favor of having as sales manager a really practical man such as myself, that it is no wonder you were stampeded into firing poor old Henderson and offering the job to me.

The whole thing, however, is profoundly disturbing to me. I want to assure you that my letter to Henderson was intended for him alone. All I was doing was merely giving him a well-deserved bawling out. I had no idea that anybody else would see this letter. And I had absolutely no intention of doing anything so base and lowdown as undermining the position of a man who is not only my boss but my friend as well. The old boy has his faults—as no one realizes more clearly than I—but he is, on the whole, a distinctly good egg. He and I have had plenty of fights, in most of which he was in the wrong, but he always treated me absolutely fair and square, and I intend to treat him the same way.

I won't take your old job. And what is more, I won't even work for the company anymore, unless you hire back Mr. Henderson at once. If you won't take both of us, you can't have either of us. And you can just take it from me that the poor old Earthworm Tractor Company would be in a sorry way indeed if it were to lose the services of both of us at the same time.

I trust that you will be able to notify me by return mail that this matter has been satisfactorily adjusted.

Yours respectfully, but firmly,
ALEXANDER BOTTS.

———

EARTHWORM TRACTOR COMPANY
EARTHWORM CITY, ILLINOIS
OFFICE OF THE SALES MANAGER

WEDNESDAY, JULY 25, 1934.

MR. ALEXANDER BOTTS,
CARE MR. OLAF ANDERSEN,
EARTHWORM TRACTOR EXHIBIT,
CENTURY OF PROGRESS EXPOSITION,
CHICAGO, ILLINOIS.

DEAR BOTTS: Mr. John Montague has turned over to me your letter of yesterday. He claims he can't understand what you are talking about most of the time, and he wants me to answer it for him. One reason for his failure to follow your arguments is that neither he nor the directors have seen your previous letter in which you suggested that you would like to be sales manager yourself. Consequently, your theory that this suggestion of yours influenced the directors in choosing you for the office is entirely unintelligible to Mr. Montague. As a matter of fact, the meeting at which you were selected as sales manager took place at ten o'clock on Saturday morning. The letter in which you nominated yourself for this office was written by you on Saturday evening. So it was obviously too late.

I am sorry that you were not present at the meeting on Saturday morning. For several weeks we had been discussing you as a possible candidate for sales manager, but many of the directors were very doubtful. They thought you were inclined to be entirely too wild and that you were hopelessly lacking in intelligence. It was to overcome these objections that I asked you to come to Earthworm City. I had hoped that if you were present, your undeniably pleasing personality might neutralize some of the opposition. And I was much disappointed when you disobeyed my instructions and failed to show up.

When the meeting took place, and you were not there, the task of pleading your cause fell entirely upon me. It turned out to be a rather difficult job. In the first place, I had to concede that most of the objections voiced by the opposition were perfectly true. I had to admit, in all honesty, that when it came to brains and intelligence you are not so very hot, and that there are times when your actions seem to indicate that you have the mind of a child of twelve. I also had to admit that you are inclined to disobey orders, that you are highly erratic and that there is a wild harum-scarum quality to your mental processes which at times seems to approach very closely to actual insanity.

In spite of all these serious drawbacks, however, I pointed out that you had certain very positive virtues. And when I got onto the subject of these virtues, I was actually surprised, myself, at the number of good qualities you seem to possess. Without stretching the truth in any way, I was able to tell the directors that you are one of the most enthusiastic and energetic salesmen we have ever had. You have complete confidence in yourself and in the Earthworm Tractor. You never know when you are licked, and you always keep going, even in the face of the most appalling difficulties. You seem to have perfect poise at all times. You are friendly and optimistic; you get along with all kinds of people; and when you start talking about tractors, you have a flow of language that is truly overwhelming. You are dependable in the sense that we know you will never lie down on the job. And, most important of all, you pass the pragmatic test—you actually sell tractors.

As soon as I had finished my talk to the directors, they all agreed with me that your high qualities of character and performance were so remarkable and so valuable that they could afford to overlook the regrettable fact that you are slightly dumb all the time and crazy part of the time. They, therefore, voted you in as sales manager. At the same time they approved the rest of our reorganization plan: Mr. John Montague, who is now president, will assume the newly created position of chairman of the board, and I will be promoted to be president of the company.

I want to thank you most sincerely for your friendship, as evidenced by your refusal to accept a promotion which you thought would be at my expense. And I want you to know that I appreciate your loyalty—even though you did not know what you were talking about when you so hastily jumped to the conclusion that I had been fired.

I trust you will wire me at once that you are accepting your new job.

Most sincerely,
GILBERT HENDERSON,
Sales Manager.

———

TELEGRAM
CHICAGO ILL JULY 26 1934
GILBERT HENDERSON
EARTHWORM TRACTOR COMPANY
EARTHWORM CITY ILL

CONGRATULATIONS ON YOUR PROMOTION STOP OF COURSE I ACCEPT THE JOB OF SALES MANAGER STOP NOW THAT THE DEPRESSION IS OVER AND NOW THAT YOU AND I ARE STEPPING INTO THESE BETTER JOBS FOR WHICH OUR TALENTS SO WELL QUALIFY US THE FUTURE IS INDEED BRIGHT STOP ADVISE WHEN YOU WANT ME TO COME DOWN AND ASSUME NEW DUTIES

ALEXANDER BOTTS

TELEGRAM

EARTHWORM CITY ILL JULY 26 1934
ALEXANDER BOTTS
CARE OLAF ANDERSEN
EARTHWORM TRACTOR EXHIBIT
CENTURY OF PROGRESS EXPOSITION
CHICAGO ILL

PLEASE REPORT HERE AT ONCE STOP YOUR FIRST JOB
AS SALES MANAGER WILL BE TO TRY TO ADJUST A
LITTLE DIFFICULTY UP AT COLD RIVER WISCONSIN STOP
DUFFIELD WATT HAS JUST SENT FRANTIC TELEGRAM
SAYING THAT PUMP WONT WORK STOP WHEN HE FIRST
TRIED TO USE IT THERE WAS A GRINDING CRASH INSIDE
AND WHEN HE OPENED THE GEAR CASE HE FOUND
IT WAS FILLED WITH SPLINTERS OF WOOD STOP I AM
DELIGHTED THAT YOU WILL HAVE TO HANDLE THIS
INSTEAD OF ME STOP THE DEPRESSION MAY BE OVER
BUT AS SALES MANAGER OF THE EARTHWORM TRACTOR
COMPANY YOUR TROUBLES ARE ONLY BEGINNING TAKE
IT FROM ONE WHO KNOWS

GILBERT HENDERSON

GOOD NEWS

ILLUSTRATED BY TONY SARG

EARTHWORM TRACTOR COMPANY
EARTHWORM CITY, ILLINOIS
INTEROFFICE COMMUNICATION

DATE: WEDNESDAY, SEPTEMBER 5, 1934.
To: GILBERT HENDERSON, PRESIDENT.
FROM: ALEXANDER BOTTS, SALES MANAGER.
SUBJECT: GOOD NEWS.

This is just a brief note to let you know that I am planning to be out of town for a while, and that I want you to look after my office work while I am away.

The cause of my sudden departure is a telegram just received from Mr. Sam Blatz, studio manager of Zadok Pictures. Inc., Hollywood, California. Mr. Blatz states that he is about to produce a motion picture called *The Tractor Man Comes Through*, with that well-known star, Buster Connolly, in the title role. The picture calls for twelve large tractors and twelve elevating graders. Mr. Blatz would like to use Earthworm machines. But they must be delivered in Hollywood within a week; otherwise he will have to employ equipment furnished by the Behemoth Tractor Company.

Naturally, I am getting after this business with all the energy I possess. It means not only a big sale but also a chance for us to get a truly stupendous amount of swell publicity. Imagine having our tractors and graders appearing in a motion picture which will be seen by millions of people all over the country, and throughout the world as well!

I am having shipped, this afternoon, twelve eighty-horsepower Earthworm tractors and twelve elevating graders on a special through-freight train which should reach California by next Monday. And, as this deal is too important to trust to our Los Angeles dealer, I am starting for the Coast tonight by plane.

You may rest assured that I am embarking on this venture with the greatest enthusiasm. After more than a month of dull office routine, I find myself with an irresistible yearning for action and excitement. And there ought to be plenty of both in Hollywood.

Don't forget to look after my job while I am gone. As it is only a little more than a month since I took your place as sales manager, on the occasion of your elevation to the presidency, you ought to remember enough about the sales work to get by all right.

As ever,
ALEXANDER BOTTS.

P.S. I wanted to tell you personally about this great opportunity. But when I called at your office a few minutes ago, your secretary informed me that, in spite of the fact that this is Wednesday, not Saturday, you had departed for an afternoon on the golf links. So I am forced to give you the glad news in this written communication, which you will probably read sometime tomorrow—provided you should happen to drift into the office.

A. Botts.

———

EARTHWORM TRACTOR COMPANY
EARTHWORM CITY, ILLINOIS
OFFICE OF GILBERT HENDERSON, PRESIDENT
THURSDAY, SEPTEMBER 6, 1934.

MR. ALEXANDER BOTTS,
CARE ZADOK PICTURES, INC.,
HOLLYWOOD, CALIFORNIA.
AIR MAIL.

DEAR BOTTS: On my arrival at the office this morning, I found your communication of yesterday awaiting me. I wish you the best of luck in your attempts to make this important sale. And I will look after your work during your absence.

However, I wish to point out, for your guidance in the future, that your place, as sales manager, is here. Actual selling work in the field should be left to our dealers and salesmen. Furthermore, I wish to remind you that it is not the policy of this company to go to the heavy expense of shipping large orders way across the country until a sale has actually gone through. And, unless you have previously consulted me, I do not want you ever again to rush off this way on the assumption that I will handle your office work for you.

Please hurry back as soon as possible.

Very sincerely,
GILBERT HENDERSON,
President.

HOLLYWOOD PLAZA HOTEL
HOLLYWOOD, CALIFORNIA

SATURDAY AFTERNOON, SEPTEMBER 8, 1934.

MR. GILBERT HENDERSON,
PRESIDENT,
EARTHWORM TRACTOR COMPANY,
EARTHWORM CITY, ILLINOIS.
AIR MAIL.

DEAR HENDERSON: Your letter is received. I want you to know that I appreciate your doubts as to the wisdom of my procedure in shipping the tractors and graders ahead of time and in coming out here myself. Also, I can understand your reluctance to take over the work of my office in addition to your own heavy duties as president of the Earthworm Tractor Company. But when I explain what I have already accomplished out here, and what I am planning to accomplish in the future, you will see that my course has been entirely justified, and you will be very glad to do your part by carrying on my work during my absence—even though it may interfere to some extent with your afternoons on the golf links.

I arrived at the Glendale Airport on Thursday morning, and at once took a taxi to the studio of Zadok Pictures, Inc., which is several miles southwest of Hollywood. My first reaction, quite naturally, was a tremendous thrill of excitement at finding myself suddenly set down right in the midst of the glamorous activities of the motion-picture business. The Zadok Studio is truly amazing—an enormous lot, half a mile square, covered with tremendous sound stages, elaborate outdoor sets, huge administrative buildings and other structures, and swarming with carpenters, electricians, various miscellaneous helpers, actors and actresses—the latter being, in some ways, the most interesting. The whole place is possessed of a quality which I can only describe as enthralling.

Even the surroundings of the studio are full of interest. The adjoining boulevard is lined with handsome filling stations, gay and colorful signboards and a lot of refreshment stands that are truly astounding—one of them, for instance, being in the form of an old mill, and another built to represent a gigantic ice-cream freezer. A short distance away is a huge alligator farm, where—for some reason which I have not as yet discovered—they are engaged in raising literally hundreds of these curious reptiles.

At one side of the studio is a low hill with a complete oil field—scores of big derricks and dozens of storage tanks. And scattered all over the landscape are countless thriving real-estate developments with clusters of little plastered bungalows sprouting up like mushrooms.

When I called at the office of Mr. Sam Blatz, the studio manager, his secretary informed me he was so busy that he could not see me until the following afternoon. This caused a certain amount of delay. But, as it turned out, it was all for the best, because it gave me a chance to roam about the studio and pick up a whole lot of firsthand information about the motion-picture business.

And when I finally saw Mr. Blatz, yesterday afternoon, I was able to speak to him in his own language, and put over a sales talk that was a real wow. Even so, I had a hard time. Mr. Blatz had already seen the Behemoth tractor people, and they had offered him unusually generous terms. However, when I explained that our machines were vastly superior, that we would agree to terms just as generous as the Behemoth and that we could make delivery early next week, his resistance broke down completely, and in less than ten minutes he closed the deal for twelve Earthworm tractors and twelve graders.

And this is not all. After thanking me most effusively for everything I had done for him, Mr. Blatz gave me a cordial invitation to remain at the studio as long as I desired. He pointed out that I could be a great help in handling the unloading of the tractors, in teaching the mechanics to operate them and, later on, in acting as technical expert on tractors during the filming of the picture. He said that he would put a studio car at my disposal, and that I would be considered a guest of the company for as long a period as I cared to remain.

This hospitable offer I promptly accepted—not, as you might suppose, because of the selfish pleasure I would derive from hanging around this fascinating studio, but rather because of the very real service I can render the Earthworm Tractor Company by remaining on the job out here a little longer. How important this service is you will realize as soon as I describe the interesting project on which I am now engaged.

After thanking Mr. Blatz, and congratulating him for his good judgment in inviting me to remain at the studio, I asked him to give me a copy of the screenplay, *The Tractor Man Comes Through*. As soon as this was in my hands, I wished Mr. Blatz a very cordial good afternoon and hurried back to the hotel in my luxurious studio car. Last night I read the play, and as soon as I had finished it, I decided that it was all wrong. Not only was it weak in a dramatic sense but it did not have anywhere near enough tractor stuff in it. What it needed was a complete revision by a real tractor expert.

Accordingly, bright and early this morning, I set to work. And I am getting along so well that I expect to have an entirely new version all ready to present to Mr. Blatz on Monday morning. The changes I am making will improve the quality of the picture so much that Mr. Blatz is almost certain to adopt all of them. And the wealth of tractor stuff which I am introducing will provide publicity of incalculable value to the Earthworm Tractor Company. A brief résumé of what I am doing will make this clear.

The original play is a rather uninspiring drama of love and hate in the swamps along the lower Mississippi River. The hero—played by Buster Connolly—is a more or less inconsequential young man in charge of a fleet of tractors which are being used in the construction of a levee. And the love interest is a girl who lives in the swamps. This is not a bad set up, but the author has spoiled it by paying too much attention to the girl and not enough attention to the tractors.

I am changing this in two ways. In the first place, I am cutting out most of the silly love passages. And in the second place, I am improving the levee-building sequences by expanding them into an exhaustive pictorial study of tractor dirt-moving operations—including not only the work of the elevating graders but also a lot of activities with dump wagons, blade graders, sheep's-foot tampers, bulldozers, scarifiers, land levelers, fresnos and a lot of miscellaneous scrapers, packers, rollers, winches, and other forms of equipment too numerous to mention.

In addition to this, I am building up the character of the hero. Instead of having him a mere second-rate straw boss on the levee, I am presenting him as a person of real consequence in the community. Besides his levee work, he is a road contractor, which gives me a chance to introduce a lot of scenes of tractors working on the roads. Also, he owns a large cotton plantation and a lumber camp, which provides an excuse for showing Earthworm tractors engaged in plowing, harrowing, cultivating and skidding logs. And, before I get through, I may even give the hero a trip up north, so we can run in some snowplow work.

Another improvement which I am making is in the method of killing off the villain. Instead of letting him drown in a very uninteresting way in the Mississippi River, I am going to have all twelve tractors run over him, one after the other. This ought to be a scene that will cause a real shudder of horror to sweep through the audience.

But my greatest and most sensational contribution is the grand climax. In the original play, when a big flood comes down the river and breaks through the levee, the hero rescues the girl from her house in the lowlands and carries her off on horseback to the safety of the hills. In my version, all this horseback foolishness is cut out, and the hero arrives in a tractor. He finds that the rescue is apparently impossible, because the girl is suffering from pneumonia or pellagra or something, and cannot be moved from her bed. In this desperate extremity, the hero hitches his tractor to the house, and drags the whole thing, with the girl in it, through miles of deadly swamps, swarming with alligators—which they can rent from the alligator farm down the road—and, after many bloodcurdling adventures, he finally reaches the safety of the hills. The hero's followers, with other tractors, haul away the barn, with the cows in it, and the various corncribs, hen houses and pigpens, with their respective contents. Not only is the girl rescued but all her property is saved. And it is just in the nick of time, because, right on the heels of this spectacular and astounding moving operation, comes the awe-inspiring influx of the swirling waters of the flood.

As you see, my improvements are going to be the making of this play, both as an artistic production and as a colossal advertisement for Earthworm tractors. So it is very fortunate that I came out here.

I must now get back to my literary labors. Good luck to you, and don't work too hard.

As ever,

ALEXANDER BOTTS.

———

TELEGRAM

EARTHWORM CITY ILL SEP 10 1934
ALEXANDER BOTTS
HOLLYWOOD PLAZA HOTEL
HOLLYWOOD CALIF

DELIGHTED THAT YOU PUT OVER SALE BUT I AM WORRIED BY YOUR STATEMENT THAT YOU AGREED TO VERY GENEROUS TERMS STOP PLEASE WRITE ME FULL DETAILS AT ONCE AND ENCLOSE COPY OF SALES ORDER STOP I DO NOT FEEL THAT THE PUBLICITY VALUE OF

THIS PICTURE WILL BE SUFFICIENT TO WARRANT YOUR
WASTING ANY MORE TIME ON IT STOP THE IMPORTANT
THING IS THE SALE OF THE TWELVE TRACTORS AND
THE TWELVE GRADERS STOP THIS HAS NOW BEEN
ACCOMPLISHED SO THERE IS NO REASON FOR YOUR
REMAINING IN CALIFORNIA STOP I AM TOO BUSY TO
LOOK AFTER YOUR JOB MUCH LONGER STOP YOU WILL
RETURN TO EARTHWORM CITY AS SOON AS POSSIBLE

GILBERT HENDERSON

HOLLYWOOD PLAZA HOTEL
HOLLYWOOD, CALIFORNIA

MONDAY EVENING, SEPTEMBER 10, 1934.

MR. GILBERT HENDERSON,
PRESIDENT,
EARTHWORM TRACTOR COMPANY,
EARTHWORM CITY, ILLINOIS.
AIR MAIL.

DEAR HENDERSON: Your telegram is here, and I am somewhat disappointed in it.

I believe you are making a mistake in feeling that the publicity this picture is going to give us is only one of the minor advantages of the deal which I have put through. I regret that you are placing undue importance on the mere selling of the twelve tractors and the twelve graders. Also, it is a bit unfortunate that you are in such a hurry to know the exact nature of the somewhat generous terms which I granted Mr. Blatz. And it is kind of too bad that you want me to send you a copy of the sales order.

As a matter of fact, the publicity is not only the most important, it is the only advantage we get out of this deal. The sale is of no consequence at all, because the terms of my agreement with Mr. Blatz provide that we lend him the tractors and graders free of charge. Hence, there has been no sale, and I can't send you a copy of the sales order because there is no sales order.

I fear that this news may be something of a shock to you, because the wording of your telegram seems to indicate that you read my former letter so carelessly that you jumped to the totally unwarranted conclusion that I had actually sold something out here. I never said anything of the kind. If you will read over my letter, you will see that I merely said I had closed a deal with Mr. Blatz. So it is not my fault if there was any misunderstanding.

However, there is nothing to worry about, because my arrangement with Mr. Blatz is perfectly fair to everyone concerned, and the terms are the best that we could get, under the circumstances. Naturally, I would have enjoyed selling all this stuff, but before I arrived, the Behemoth Tractor Company had already offered to supply, absolutely free of charge, twelve of their tractors and twelve of their elevating graders. And Mr. Blatz, although he preferred Earthworms, did not want them badly enough to actually pay any money for them. So there was only one thing to do. As a sale was clearly impossible, I abandoned all thought of making it, and concentrated my efforts on this wonderful opportunity for motion-picture publicity. And I have succeeded admirably. All I had to do, in addition to lending them a mere hundred thousand dollars' worth of tractors and graders, was to sign a simple agreement releasing the motion-picture company from liability in case there is any damage to this property while it is in their possession. And in return for this, we get a chance to put over at least a million dollars' worth of splendid publicity. So, you see, the advantages are all on our side.

However, if we are going to exploit this magnificent opportunity to the fullest possible extent, it will be necessary for me to stay out here long enough to make sure that the picture is loaded with as much Earthworm-tractor propaganda as it can possibly hold. In a former letter I explained how I was revising the screen play so that the picture will do right by our tractors. The revision was completed yesterday, and the next step is to submit it to Mr. Blatz. But I cannot do this right away, because the guy is out of town. He left unexpectedly last night by airplane for New York, where he is to confer with Colonel Zadok, the head of the company, and with various bankers who seem to have acquired, in some mysterious and insidious manner, a considerable influence in the business. Mr. Blatz expects to be back on Wednesday, and he left word that he hoped I would be able to attend a story conference which he will hold on that day for the purpose of discussing the tractor picture. Obviously, there is only one course for me to pursue. I will have to stick around here until Wednesday, at least.

In the meantime, I am finding plenty to occupy me. This morning, with the help of a large force of expert mechanics from the production department, I unloaded the tractors and graders, which had arrived on a nearby siding, and brought them over to the lot.

This afternoon I visited a number of people who are to be connected with the forthcoming tractor picture, and tried to sell them my ideas for revising the script. I wanted to get them on my side, so that they would back me up when I present my plans to Mr. Sam Blatz at the story conference on Wednesday. My efforts, however, were not very successful.

The first person I interviewed was Mr. Buster Connolly, the star of the picture. I found him a short distance outside the lot, looking at a large and elaborate bungalow which belongs to him and which was in the process of being moved to the Zadok studio from another studio where he had previously worked. It appears that every motion-picture star that amounts to anything has to have a bungalow where he can loll around at such times as he is not actually working. When I arrived upon the scene, Mr. Connolly's bungalow was coming past the oil derricks on the shoulder of the hill at one side of the Zadok lot. It was being dragged along very slowly and painfully by means of an antique and clumsy arrangement of cables, pulleys and winches.

This gave me a wonderful opening. After explaining the changes I wanted to make in the forthcoming picture, I offered to bring up my tractors and put on a real demonstration. I told Mr. Connolly that I would move his bungalow down to the lot so fast that it would make his hair stand on end, and would also give him an idea of the sensational effect we would produce with the spectacular house-moving climax which I had decided to put into the picture. But Mr. Connolly was not impressed.

"I won't let you touch my bungalow," he said. "And, furthermore, I don't like your machines. When Sam Blatz gets back I'm going to see him and have practically all the tractor scenes taken out of the picture."

"But, Mr. Connolly," I said, "the tractors will be the making of the picture. They'll be the most interesting thing in it."

"That's just the trouble," he retorted angrily. "The audience will be looking at the tractors all the time, and they won't pay any attention to my acting. Thus, the picture will be ruined."

"Oh, I see," I said. And, indeed, I began to see only too well what was on his mind. He knew the tractors would be splendid picture material. He was afraid they would steal the show from him. And he was so stubborn that I found it impossible to argue with him. Consequently, after a very disappointing interview, I left him and went around to see the director of the picture.

The director sprung an idea that was exactly the opposite of Mr. Connolly's, but just as bad from my point of view. He said he was in favor of cutting out practically all the scenes involving tractors, because tractors, in his opinion, are "slow-moving props," without any picture value whatsoever. I offered to disprove this assertion by giving him a demonstration. He refused. Then I tried to describe the thrilling effects which could be obtained by showing our Earthworms plowing majestically across horrible swamps, through dense forests and over rugged mountains. But he would not listen. He had seen the tractors coming into the lot that morning. They moved slowly. Hence, they were slow-moving props, entirely devoid of interest. And nothing I could say had any effect. So, finally, I had to abandon my efforts, and leave him in the same state of besotted ignorance in which I had found him.

My next, and final, call was on the production manager, who has charge of the studio equipment. This guy promptly announced that he was going to advise Mr. Blatz to cut out all the tactor scenes, on the ground that they would necessitate very expensive sets, or even more expensive trips to outside locations. And he was just as opinionated as Mr. Connolly and the director, so I soon gave up all thought of reasoning with him, and walked away in disgust.

But don't get the idea that I am discouraged. Mr. Sam Blatz is the boss around here. He has already shown his intelligence, and his appreciation of tractors, by arranging with me for the use of our machines. Consequently, I am able to contemplate the future with the greatest of confidence. When I see Mr. Blatz at the story conference on Wednesday, I will be prepared to present my ideas in such a clear, forceful and convincing manner that he is almost certain to follow my advice. Hence, the ignorant opposition of such unimportant underlings as the star, the director and the studio manager will count for nothing at all. And the picture will be produced in such a way that it will be not only a credit to Zadok Pictures, Inc., but also an invaluable piece of propaganda for the Earthworm Tractor Company.

Yours enthusiastically,
ALEXANDER BOTTS.

TELEGRAM
EARTHWORM CITY ILL SEP 13 1934
ALEXANDER BOTTS
HOLLYWOOD PLAZA HOTEL
HOLLYWOOD CALIF

OUR ATTORNEY BELIEVES THAT YOUR LENDING
TRACTORS TO MOTION PICTURE COMPANY IS VIOLATION
OF NRA CODE ALTHOUGH AS USUAL HE IS NOT QUITE
SURE STOP IN ANY EVENT IT IS ABSOLUTELY CONTRARY
TO OUR POLICY STOP YOU WILL NOTIFY MR BLATZ THAT
HE MUST BUY THE TRACTORS AND GRADERS AND PAY
FOR THEM STOP IF HE REFUSES YOU WILL REMOVE
ALL OF THIS MACHINERY AND TURN IT OVER ON
CONSIGNMENT TO OUR LOS ANGELES DEALER IN HOPES
OF FUTURE SALE ELSEWHERE STOP YOU WILL RETURN
TO EARTHWORM CITY AS SOON AS POSSIBLE AS WORK
IN YOUR OFFICE HERE IS PILING UP
GILBERT HENDERSON

———

Hollywood Receiving Hospital
Hollywood, California

Thursday Evening, September 13, 1934.

Mr. Gilbert Henderson,
President,
Earthworm Tractor Company,
Earthworm City, Illinois.
Air Mail.

DEAR HENDERSON: It is my painful duty to report to you that I have
been suddenly overwhelmed by a series of crushing and incredible disas-
ters. Your telegram has been forwarded to me here at my new address, but,
in the present situation, it does not mean anything at all. I cannot do any
of the things you want me to do. And I cannot even do any of the things I
want to do myself. I cannot stay out here to help with the tractor picture,

because there is not going to be any tractor picture. I cannot hurry back to Earthworm City, because I am laid up in the hospital with a broken ankle. And I cannot repossess the tractors and graders, because there are no tractors and graders any more. They have, in fact, ceased to exist as such. So, about the only thing I can do is sit up in bed and write you an account of the unbelievable combination of catastrophes which have defeated all my best-laid plans.

The first blow descended upon me early yesterday evening at the story conference held in Mr. Blatz's office at the studio. All the people concerned in the tractor picture were there, and each one was prepared to present his own ideas on the subject. But the wind was immediately taken out of all our sails by Mr. Blatz himself, who had just arrived from New York full of a whole set of new and outlandish ideas.

Mr. Blatz told us that the financial powers had decided that what the company needs at the present time is something that will ring the cash register in a big way. And they had concluded that the best means of accomplishing this was to produce a picture which would cost more than any other picture heretofore produced in the entire history of the industry. The idea, if you can call it that, seems to be that, instead of making several ordinary pictures which would cost two or three hundred thousand dollars apiece and bring in perhaps a half a million apiece, they will concentrate all their efforts on one production which will cost over a million dollars, and will bring in—they hope—several millions at the box offices.

In accordance with this plan, they are going to junk the tractor drama and three or four others that had been scheduled for early production, and put everything they have into a stupendous, monumental, sensational, magnificent, and overpowering epic super-spectacle based on Milton's *Paradise Lost*.

Mr. Blatz admitted, quite frankly, that, from a picture point of view, Mr. Milton's stuff is not so hot, but he was fairly goggle-eyed with enthusiasm over the commercial possibilities. He pointed out that this man Milton has been built up and advertised so long and so extensively by all the college professors and all the high-school teachers of English throughout the entire country that his name has become a household word, synonymous with culture and the higher things of life. This situation is just naturally made to order for the publicity department, which will rush about the country organizing John Milton Booster Clubs in every city, town, hamlet and crossroads. Indorsements will be sought, and doubtless received, from women's clubs, parent-teacher associations, literary societies and all the rest of the innumerable organizations which are interested in improving the mind and raising the standard of good taste and artistic appreciation in America. Fathers, mothers, clergymen, school teachers, college professors and civic leaders generally will be so thoroughly sold on the high value of the picture that they will urge every man, woman and child throughout the length and breadth of the land to see it.

As director of the picture, they are hiring a world-renowned Russian theatrical producer. This bird knows nothing about motion pictures, and he cannot even speak English, so he will be completely useless around the studio. But the real work can be done by an American assistant director who really knows his business, and the name of the famous Russian will tend to pull in that large and influential group of Americans who believe that nothing can be truly worthwhile in an artistic way unless it comes out of Europe.

And this is not all. It is realized that there exists in America a considerable substratum of ignorant dumbbells who have a deep suspicion of anything they think is highbrow. To attract this element, they will use, in the cast, a whole group of stars who have a known appeal to the lower classes. Besides Buster Connolly, who has a great following among lovers of thrills and Western melodrama, there will be four popular comedians, three lady stars of the hot-mamma type, and various handsome juvenile men, to say nothing of acrobats, adagio dancers, circus performers, stunt men and an enormous corps de ballet. The sets and the costumes will be the most elaborate ever seen—they have already ordered one thousand harps and one thousand pairs of wings, with real feathers, for the angels—and the dance ensembles will be lavish beyond description. In fact, there is only one place where they will economize—they won't have to pay anything to the original author.

As soon as Mr. Blatz had completed a description of his ambitious plans, the entire group at the conference—with one notable exception—gave voice to their admiration and approval. The only sour note was sounded by myself, in the form of a plaintive protest at the way they were letting the Earthworm Tractor Company down with nothing to show for all the expense we had incurred in bringing the tractors to California at Mr. Blatz's express request. But my objections were completely drowned out in the great flood of enthusiasm for Mr. John Milton.

And then, while I was still reeling from the effects of this first blow, there descended upon my unfortunate head a second disaster of even greater magnitude. The building was suddenly shaken by the shock of a distant explosion, and soon after we began to hear cries of "Fire!" We all rushed outside, and followed a crowd of excited studio employees to the western boundary of the lot. And here a truly appalling sight met our eyes.

One of the oil-storage tanks up among the derricks on the nearby hill had caught fire. Apparently, this tank was only partly filled with oil; it must have contained a certain amount of air mixed with oil vapor in deadly proportions. At any rate, it had exploded, and spread the fire all over the hill. The flames were roaring up from dozens of derricks and tanks. And, as we watched, another tank blew up with a tremendous report, and a great river of burning oil started down the hill toward the studio of Zadok Pictures.

Much to our relief, the flow of this fiery river was arrested, high up on the hill, by a long, low bank of earth, which had presumably been thrown up as a protection for the studio in case of just such an emergency as this. But our relief was short-lived, for it soon appeared that this embankment was not high enough. A small quantity of the burning oil came over the top, and it was evident that, if a few more tanks let loose, the blazing fluid would come pouring down the hill and engulf the studio. Fire engines were already on the way from Culver City, Los Angeles, Hollywood, Beverly Hills and Santa Monica. But it was obvious that they would be of little use.

There was only one thing to do. I rushed over to the shed where my tractors and graders were stored. And, as I went, I called loudly for my mechanics to man the machines. Fortunately, several companies were on the lot, taking night scenes. So most of my mechanics were present. And they—gallant and intelligent fellows that they are—had already conceived the same idea which had come to me.

In less than five minutes I had men on every tractor and every grader. And in less than ten minutes, we came charging out through the big side gate of the lot, and advanced up the hill toward the fire.

By this time, the electrical department had mounted hundreds of enormous klieg lights at various vantage points. And these, in conjunction with the light from the fire, made the entire scene as bright as day.

Directly in front of us on the slope of the hill stood Buster Connolly's expensive but silly-looking bungalow. We promptly hooked onto it with three tractors and hauled it down into the lot. Then I started running the twelve tractors, each one pulling a grader, back and forth across the face of the hill so as to throw up a really effective barrier about halfway between the studio and the small dam which was temporarily holding the burning oil in check.

It began to look as if the studio might be saved.

It began to look as if the studio might be saved. The tractors were roaring on their way, and the elevating graders were plowing up the earth and casting it onto the new embankment so fast that victory would be assured in fifteen or twenty minutes.

At this juncture, however, my own participation in the fight was abruptly terminated. As I was rushing to and fro, and hither and yon, shouting orders and encouragement to my trusty followers, I inadvertently tripped over a large plant of the variety known as prickly pear. This unfortunate accident not only filled me up with a lot of spines, in a most annoying way, but also caused me to sustain, as I landed on the hard ground, a broken ankle. It was the same ankle which I injured once before down in Mississippi. A number of husky lads promptly picked me up and carried me down to the highway. And before I knew exactly what was happening to me, I had been loaded into an ambulance and dragged away to this hospital, where the doctors proceeded to give me ether, so they could properly set the broken bones.

When I finally regained the full use of my mental faculties, several hours later, I insisted that the nurse call up the studio and find out how things were going. The news that she brought me was partly good but mostly bad. When the inevitable break in the upper embankment occurred, the lower one, so hastily constructed by our tractors and grad-ers, had reached a height sufficient to stem the tide of the burning oil. So the entire Zadok Studio had been saved. But at the time of the break, our tractors and graders were on the upper side of the new earthwork. The brave mechanics had time to leap to safety over the barrier, but the twelve tractors and the twelve graders were all left behind, and were completely destroyed.

Thus ends the magnificent enterprise upon which I embarked, last week, with so much hope and enthusiasm. At the moment, I am feeling too wretched and miserable to make any definite plans for the future. But, probably, as soon as I am well enough, I shall come creeping piteously back to Earthworm City, hoping and praying that my job has not been taken away from me.

Yours, with deep sorrow,

ALEXANDER BOTTS.

HOLLYWOOD RECEIVING HOSPITAL
HOLLYWOOD, CALIFORNIA

FRIDAY AFTERNOON,
SEPTEMBER 14, 1934.

MR. GILBERT HENDERSON,
PRESIDENT,
EARTHWORM TRACTOR COMPANY,
EARTHWORM CITY, ILLINOIS.
AIR MAIL.

DEAR HENDERSON: I have a little news for you. My ankle is getting along very well, and I am feeling quite comfortable.

This morning I got in touch with the local adjuster for the Illinois Eureka Fire Insurance Company, which, as you know, handles our busi-ness. The adjuster has inspected the remains of our tractors and graders, and is reporting them as a total loss. So, under the terms of our policy, we are to receive the full value, plus freight. This means that the fire wasn't such a bad idea after all; we get paid for these machines just the same as if we had actually sold them.

And here is more news. Mr. Sam Blatz called on me this morning to thank me for my part in saving the studio and thus making it possible for them to go ahead, without any delay, in the production of their great super-spectacle. When I asked if he was referring to the *Paradise Lost* picture, he replied that he was, except that their plans had been slightly changed and they were not going to use the great Milton epic after all. It appears they had hired some guy to read the book for them, so they could find out

As I was rushing to and fro, and hither and yon, shouting orders and encouragement to my trusty followers, I inadvertently tripped over a large plant of the variety known as prickly pear. This unfortunate accident not only filled me up with a lot of spines, but also caused me to sustain a broken ankle.

what it is all about. And they had also consulted with the Hays office. Most of the authorities had then agreed that it would be pretty difficult to get *Paradise Lost* past the censors and the various purity organizations. Apparently, one of the main characters in the book is Satan himself, and this puts entirely too much emphasis on sin. So they have decided to do John Bunyan's *Pilgrim's Progress* instead. This book—which apparently has nothing to do with Paul Bunyan, owner of the famous blue ox— contains a number of objectionable passages, but they can be eliminated without much trouble.

Mr. Blatz pointed out that the switch to the Bunyan opus was in reality only a minor change in their plans. The really essential features—such as the elaborate production and the publicity aimed at the culture groups—will go through just the same. The only real difference would be that, instead of forming Milton societies all over the country, the publicity department would bend its energies toward the formation of bigger and better Bunyan clubs.

"And in this new set up," Mr. Blatz explained, "we find we won't need Buster Connolly, so we are going to have him make the tractor picture after all—only there will be a lot of changes in that too."

"I suppose you're going to leave out all the tractors?" I asked suspiciously.

"I suppose you're going to leave out all the tractors?"

"I'm afraid we can't," said Mr. Blatz. "You see, the most spectacular part of the picture has been made already."

"I don't understand."

"It's like this, Mr. Botts. When that fire started, we had four separate companies working on the lot. And the entire production force was present, all ready to take advantage of this great opportunity. The electricians set up their lights, and the directors and the cameramen went into action. And the results are really extraordinary. Never before has a fire of this size been covered by so many expert cameramen with such a wealth of high-grade equipment. And never before have I seen such beautifully taken fire pictures. We have the whole thing—blazing oil derricks, exploding tanks, the moving of that bungalow from out of the very jaws of death, the building of the embankment, the escape of the tractor operators in front of the blazing river of oil and the final destruction of the tractors and the graders. We even have a delightful bit of comedy relief—where some jackass tripped and did a beautiful high gruesome into a prickly pear. It is all so astounding, so magnificent, so colossal, that we are going to use it as the climax of our tractor picture. To do this, we shall have to shift the scene of our story from the Mississippi swamps to California, and have the hero rescue the girl and her house from an oil fire. We want to make this a real tractor picture, with lots of tractor stuff all through it, so we are hoping that you will be able to lend us twelve more tractors and twelve more graders to be used in the earlier sequences."

"Nothing would please me more!" I said. And then my warm enthusiasm was suddenly chilled by the icy fingers of a cold doubt. "I am afraid, however," I said sadly, "that it can't be done. Look at this."

I handed him your telegram. He read it and glanced at me inquiringly.

"We must face the facts," I said. "You will have to buy these tractors and graders, and pay real money for them, or else you'll have to use Behemoth machines.

"We can't use Behemoths," said Mr. Blatz, "because the most important part of the picture has already been made with Earthworms." He pondered the matter for some time. "Oh, well," he said at length. "It's only a hundred thousand dollars—a mere drop in the bucket as compared to what we ought to take in on this picture. Do you want me to sign an order or something for these things?"

"As a matter of form, you might as well," I said nonchalantly. "And you might make me out a check also."

So he did, and I am enclosing the order and the check herewith. Please ship the stuff at once. And I'm afraid you'll have to handle my office work a little longer, because the state of my ankle makes it necessary for me to leave tomorrow for several weeks' vacation, as the guest of Mr. Sam Blatz at his luxurious ranch in Hidden Valley up in Ventura County.

Yours respectfully,
ALEXANDER BOTTS.

HOLLYWOOD IS
WONDERFUL, BUT–

ILLUSTRATED BY TONY SARG

EARTHWORM TRACTOR COMPANY
EARTHWORM CITY, ILLINOIS
OFFICE OF GILBERT HENDERSON, PRESIDENT

MONDAY, OCTOBER 1, 1934.

MR. ALEXANDER BOTTS,
CARE MR. SAM BLATZ, STUDIO MANAGER,
ZADOK PICTURES, INC.,
HOLLYWOOD, CALIFORNIA.
AIR MAIL.

DEAR BOTTS: Two weeks ago you wrote that you had completed your sale of tractors and graders to Zadok Pictures, Inc., and that you were taking a short vacation to recuperate from an injured ankle. Since then we have had no news from you at all.

We are very anxious to have you return to Earthworm City at once. The Earthworm Tractor Company has just taken over the Crowder Shovel Corporation, of Chicago, and we are going to market their entire line of heavy excavating machinery under the name Earthworm. We want you here at the home office, so you can work out plans for pushing this new group of products.

Please advise when we may expect you.

Very sincerely,
GILBERT HENDERSON,
President.

HOLLYWOOD PLAZA HOTEL
HOLLYWOOD, CALIFORNIA

WEDNESDAY, OCTOBER 3, 1934.

MR. GILBERT HENDERSON,
PRESIDENT,
EARTHWORM TRACTOR COMPANY,
EARTHWORM CITY, ILLINOIS.
AIR MAIL.

DEAR HENDERSON: Your letter is received, and I hasten to report that I am getting along beautifully. My ankle, which was not hurt as badly as I had supposed, is now practically well. I have been spending the past few days at the Zadok Studio, advising them on how they should handle their recently purchased Earthworm tractors in their forthcoming tractor picture. I want to be sure they get the greatest possible dramatic effect out of the machines, and also—although I did not tell them so—I want to be sure they give us as much publicity and free advertising as possible. This advisory work is now completed, and I was just on the point of starting back to Earthworm City to take up once more my regular duties as sales manager.

But now that I have received your letter asking me to hurry home, I have decided that I really ought to stay here a little longer. This does not mean that I am disregarding your instructions. It means, on the contrary, that I am planning to carry out your desires in a way that is far superior to anything you could have even thought of.

What you want is to launch the Earthworm Tractor Company into the power-shovel business with as large and sensational a splash as possible. For this we shall need lots of publicity. And I have decided that the best and cheapest way for us to achieve this publicity is for me to stay right on the job out here and devote all my energies to the task of putting our shovels into the movies, so that they may be shown, in all their glory, on the screens of all the motion-picture theaters of the country.

Now, it is obvious that I am better fitted than anyone else in our organization for carrying out this important project. In the few weeks that I have been in Hollywood I have acquired a thorough knowledge of the inside workings of the motion-picture business. I am personally acquainted with Mr. Sam Blatz, general manager of the great Zadok Studio. And, having already succeeded in placing a group of our tractors in one motion picture, I know exactly how to go about placing a group of our shovels in another.

I am planning an attack that will be simple, direct and powerful. I will spend several days working out and writing up a fast-moving, red-hot, dramatic story dealing with the life of a power-shovel operator. As soon as it is completed, I will offer it, absolutely free of charge, to Mr. Sam Blatz. As he is a man of unusually high intelligence, he will at once recognize its possibilities, and seize upon it with the greatest eagerness. Then, when he starts producing the picture, he will find that the story calls for several shovels, and we shall make a nice little sale—which is not to be scorned, even though our main idea is the publicity.

In a very short time I hope to send you news of my complete success.

As ever,
ALEXANDER BOTTS.

———

TELEGRAM
EARTHWORM CITY ILL OCT 5 1934
MR ALEXANDER BOTTS
HOLLYWOOD PLAZA HOTEL
HOLLYWOOD CAL

I WISH TO REPEAT MOST EMPHATICALLY THAT A SALES MANAGER BELONGS AT THE HOME OFFICE JUST AS THE GENERAL OF AN ARMY BELONGS AT HEADQUARTERS AND NOT IN THE FRONT LINE TRENCHES STOP PLEASE RETURN TO EARTHWORM CITY AT ONCE STOP OUR LOS ANGELES DEALER CAN HANDLE THE MOTION PICTURE SHOVEL DEAL STOP IN VIEW OF THE FACT THAT WE DO NOT YET KNOW WHETHER OR NOT MOTION PICTURE PUBLICITY WILL BE OF ANY REAL VALUE TO US I DO NOT FEEL THAT YOU SHOULD WASTE ANY MORE TIME ON THE MATTER

GILBERT HENDERSON

HOLLYWOOD PLAZA HOTEL
HOLLYWOOD, CALIFORNIA

FRIDAY EVENING, OCTOBER 5, 1934.

MR. GILBERT HENDERSON,
PRESIDENT,
EARTHWORM TRACTOR COMPANY,
EARTHWORM CITY, ILLINOIS.
AIR MAIL.

DEAR HENDERSON: Your telegram arrived this morning, and I was much flattered to learn how indispensable you consider my presence at Earthworm City. I was also much pleased at the apt way in which you compare my position as sales manager to that of a general in command of an army. I want to assure you that I agree with you entirely in your estimate of my importance to the company. But I cannot help feeling that you are a little far-fetched in your ideas on the subject of headquarters versus front-line trenches. This is entirely a matter of temperament and circumstances. I will admit that it was entirely correct, in the late war, for General Pershing to sit around at Chaumont like a spider in the center of his far-flung web of communications. Under the circumstances, it was doubtless the most effective procedure that he could have used. And it succeeded. But there have been other wars, in which other leaders have succeeded by using entirely different methods. In my present campaign, I feel that it will be the part of wisdom to employ the sort of tactics which are best adapted to my own personality. This means that I will not attempt to play any part so foreign to my nature as that of a spider sitting in the center of his web. It is far better that I should ride forth, like Joan of Arc, at the very front of the attack.

This procedure is especially desirable at the present moment because my campaign is not going as well as I had hoped, and it is, therefore, necessary that the commander in chief be present in person to direct the operations. Although my plans have received a temporary setback, I have no doubts regarding my ultimate success. And in order that you may understand the exact situation out here, I will give you a brief account of what I have done and what I am planning for the future. I hope you will read what I have to say with great care, because I want you to realize that my selling campaign here has been planned and executed with unusual efficiency and thoroughness, and my failure to accomplish any results so far has been due entirely to circumstances outside my control.

I spent three whole days—Monday, Tuesday and Wednesday—in creating, elaborating, embellishing and setting down what is probably one of the most remarkable motion-picture scenarios ever produced in the entire history of the industry. The hero of this opus, who is a steam-shovel operator, is one of nature's noblemen—a strong, courageous, red-blooded, 100 percent American he-man. The heroine is young, good and beautiful, and I am making her just as fascinating and voluptuous as I can without getting into trouble with the censors. And the villain, who is probably the most striking character in the play, is a combination of the worst qualities of all the most objectionable people I have ever met in my entire career as a tractor man—which covers a lot of ground. He is so utterly and unspeakably low, vile and repulsive that—if he really existed—you could not even pass him on the street without wishing to put your foot in his face. He is sure to go over big—especially with those more discriminating people who like movies which are sophisticated, yet delicately handled.

The personalities of the three main characters are skillfully developed and presented in a series of colorful incidents, climaxes and super-climaxes which not only provide drama of the highest order but also give our power shovels an opportunity to show off in a way that will convince everyone of their superior quality. The play ends in a blaze of glory with the hero rescuing the heroine from almost certain death, and using his trusty machine to shovel the villain over the edge of a mighty precipice.

Taking it all in all, the scenario is a lollapaloosa. I presented it to Mr. Blatz at the Zadok Studio yesterday morning, and, after skimming through it, he agreed absolutely with my opinion that I had produced something which might well be described as remarkable, not to say astounding. Unfortunately, he also told me, gently but with sickening firmness, that he did not feel justified in producing my picture at the present time. He stated that he was already making the tractor picture, and that he did not want to risk another machinery drama until he found out whether the first one would be successful. Naturally, I disagreed with him with the greatest vehemence. But after arguing for about half an hour, I saw that he was—as we writers would say—adamant. I therefore withdrew as gracefully and politely as I could, and spent the rest of the day in futile assaults on two other major studios.

At neither of these other places was I able to reach the higher executives, so I had to waste my efforts on a couple of incredibly stupid, unimaginative fellows known as story editors.

These people used some of the same arguments which I had heard in my first attempts to introduce Earthworm tractors at the Zadok Studio. They said that shovels are dull, uninteresting, slow-moving props, entirely lacking in entertainment value. They would not listen to me when I offered to prove, by an actual demonstration, that they were wrong. And they would not even read my scenario. So, when I returned to the hotel last night, I will have to admit that my heart was heavy with discouragement.

I did not, however, remain in this state of mind for long. A brief analysis of the situation convinced me that the only thing standing in the way of my complete success was my inability to persuade these motion-picture people to come to a shovel demonstration, so that I could overcome their abysmal ignorance regarding the pictorial possibilities of our splendid machines. And this diagnosis of the difficulty at once suggested a happy remedy. If the movie moguls would not come to me, I would take my demonstration to them. And I would present it in the medium of expression with which they are most familiar—in other words, I would make a short motion picture, showing one of our shovels performing the thrilling and spectacular maneuvers which are called for in the climax of my scenario, and I would then take the film around and run it off before the astonished and delighted eyes of the various studio executives, thus raising them to such a high pitch of enthusiasm that my only remaining task would be to decide which one of them should be the lucky man with whom I would consent to do business.

After deciding on the above promising plan of campaign, I turned in for a good night's rest. And this morning I began to put my ideas into action.

First of all, I visited the machinery house which has the Southern California agency for Crowder shovels—now Earthworm shovels. In answer to my request for the loan of a machine, they stated that they had none in stock, and that there was only one in the entire territory—a two-yard gasoline model owned by a local contractor named Joe Blake.

Note: Apparently, this agency has not been pushing these shovels as hard as they should. I will get after them hot and heavy as soon as I have time. This morning, unfortunately, I had too many other things to do.

I at once proceeded to call on Mr. Joe Blake, the contractor. He informed me that his Crowder—now Earthworm—shovel is up in the mountains near Arrowhead Lake, where he is preparing to build a dam in the bottom of a rather inaccessible canyon. His description of the location indicates that it will be admirably suited to my purposes. The machine is on the upper rim of the canyon, getting ready to cut a zigzag road down the steep side wall into the trackless depths below—this road being a necessary preliminary, so that he can get his equipment and supplies to the site of the dam. When I asked Mr. Blake if he would permit me to take some motion pictures of his shovel, he replied that he would be delighted, and that he would drive me out next Monday.

Having accepted this kind offer, I rushed around to a small independent studio and arranged to hire a cameraman with a motion-picture camera and plenty of film, and also a reasonably competent actor and an actress. In addition, I have commissioned a painter to make a big "Earthworm" sign to put on the shovel. And I have purchased from a department store two clothing dummies, one male and one female.

I am using these dummies because I don't want to kill anybody. They will be substituted for the real flesh-and-blood players in the scenes where they are picked up and whisked around by the great shovel.

With any kind of luck at all, I ought to be able to complete my picture in one day—next Monday. But it will probably take a couple of days more to develop and exhibit the film. So the good news of my success, although it is inevitable, will probably not reach you till the latter part of the week.

As ever,
ALEXANDER BOTTS.

———

TELEGRAM
EARTHWORM CITY ILL OCT 8 1934
MR ALEXANDER BOTTS
HOLLYWOOD PLAZA HOTEL
HOLLYWOOD CAL

I AM GETTING TIRED SENDING OUT MESSAGES TO WHICH YOU PAY NO ATTENTION STOP I WISH TO REMIND YOU THAT A COUPLE OF MONTHS AGO YOU SIGNED A ONE YEAR CONTRACT TO WORK FOR THE EARTHWORM TRACTOR COMPANY AS SALES MANAGER STOP WE INSIST THAT YOU RETURN TO EARTHWORM

CITY AT ONCE AND FULFILL THE TERMS OF THIS
CONTRACT BY HANDLING THE WORK FOR WHICH YOU
WERE HIRED

GILBERT HENDERSON

———

HOLLYWOOD PLAZA HOTEL
HOLLYWOOD, CALIFORNIA

MONDAY EVENING, OCTOBER 8, 1934.

MR. GILBERT HENDERSON,
PRESIDENT,
EARTHWORM TRACTOR COMPANY,
EARTHWORM CITY, ILLINOIS.

DEAR HENDERSON: Your telegram is here, and in some ways I am
almost beginning to wish that I had heeded your advice and come back
to Earthworm City before I even started in on this power-shovel-motion-
picture project. At the moment, nothing would please me more than to
follow your instructions by hopping on the fastest airplane I could find
and heading for Illinois. Such a course of action, however, would, under
the present circumstances, be distinctly cowardly and dishonorable. The
situation out here has suddenly become so involved and so difficult that I
feel it is my duty to remain a few days longer for the purpose of attempting
some adjustment in the various lawsuits and claims for damages which
seem to be coming in from all directions.

The unfortunate state of affairs in which I find myself this evening is
rendered doubly distressing by the fact that it arrived so suddenly. When I
awoke this morning, my mind was calm, serene and completely free from
worry. And this happy condition continued for quite a while. My breakfast at the hotel was excellent. Soon afterward, Mr. Blake, the contractor,
arrived with his car, and we started for the mountains near Arrowhead
Lake, taking with us the big motion-picture camera, the cameraman, several cans of film, the actor and the actress, the male and female clothing
dummies and the big Earthworm sign.

Toward the end of the morning, after a delightful drive, we stopped for
a bite to eat at a large vacation camp in a beautiful evergreen forest about

half a mile distant from the point on the rim of the canyon where Mr. Blake's shovel was located. There were quite a number of people staying at this camp, having been attracted there by a series of important tennis matches. But as soon as the crowd discovered that we were about to take some exciting motion pictures, most of them deserted the tennis courts and followed our car over to the edge of the canyon. This crowd proved to be a good deal of a nuisance; we had to keep shooing them back out of the way all the time. But they were also a source of considerable gratification to me; for it was evident, from the eager way in which they pressed forward, and from the frequent bursts of applause and cheering, that we were putting on an act which had real audience appeal.

It did not take me long to get my performance under way. I located a medium-sized but very sturdy tree which grew on the rim of the canyon in such a way that one of its limbs extended far out over the edge. At this point the canyon wall sloped down, at an angle considerably steeper than forty-five degrees, to the stream several hundred feet below. My scenario called for an absolutely sheer cliff, but I was glad to modify this requirement slightly, especially in view of the fact that I was going to have the young actress climb the tree, and I did not want to run the risk of losing her entirely in case she made a slip. And it really made no great difference, as the cameraman assured me that, by skillful placing of the camera, he could create the illusion of a truly appalling precipice. As soon as these preliminaries had been decided, the actual taking of the picture got under way, the various scenes progressing in the following order:

The beautiful heroine—played by the young actress whom I had hired—comes tripping lightly through the forest on the way to meet her lover, who is supposed to be operating his big power shovel not far away. In her dainty hand she bears the map which shows where the treasure is buried. Suddenly she meets the villain—played by the actor whom I had brought along. His greedy eyes recognize the map. He attempts to seize it. The heroine flees shrieking through the forest; the villain pursues; and we have a really swell chase sequence.

Finally, the heroine reaches the edge of the canyon. She cannot go on. But she cannot go back, either, for the villain is right after her—so close that she can almost feel his hot breath on the back of her neck. She is, in short, in something of a dilemma. But all at once she spies the tree—the one which I had previously selected. She climbs the tree, and finally reaches the end of the branch which extends far out over the yawning chasm.

The villain gazes up at her, and a leering smile spreads over his repulsive face. He has not exactly caught the lady as yet, but he certainly has her out on a limb. And if he can only succeed in dashing her into the canyon, he can climb down later on and abstract the coveted map from her lifeless fingers. With a snarl of rage, he pulls a hatchet from his belt and starts chopping down the tree.

At this moment the excitement and suspense reach a very high point. Suddenly, in the distance, appears the handsome hero—played by myself—driving through the forest in the huge shovel which bears the recently attached sign, Earthworm. The hero sees that the heroine needs his help. He puts on full speed. The great machine fairly flies across the ground. The motor roars. But the villain is so absorbed in his chopping that he does not hear anything until the shovel is almost on top of him. At this point the grand climax is reached.

But before I relate exactly what happened, I wish to pause a moment to explain that right here I had intended to stop the camera and substitute the two clothing dummies for the two real people who were playing the parts of the heroine and the villain. This change seemed advisable because from here on—as I have previously indicated—the scenario called for some pretty rough action. Just as the villain raised his hatchet for the final blow which would sever the trunk of the tree and send the heroine to a cruel death on the rocks far below, I had planned to have the hero—played by myself—give a mighty lunge with the huge dipper, scoop up the villain and shovel him, like so much refuse, far out over the edge of the precipice. Then, with a skillful swing of the boom, I was going to bring the dipper up under the young lady just as her weary fingers were supposedly losing their grip on the branch, and pluck her from the tree like a ripe plum. I would then land her safe and sound on terra firma, and there would be a final close-up showing the two lovers locked in a fond embrace.

That, as I say, was my plan. And it is easy to see what a wonderful picture we would have had, provided everything had worked out as I had expected. But at the critical moment there occurred a small accident which ruined the whole thing.

As I sat in the driver's seat of the mighty shovel, steering it skillfully and accurately in its awe-inspiring progress along the rim of the canyon toward the tree in which the heroine was so precariously perched, I suddenly noticed that the machine was beginning to tilt at a very alarming angle. I looked around, and it was not long before I realized what was happening. I had driven so close to the canyon wall that the tremendous weight of the shovel had caused the earth to start crumbling away. Quick as a flash, I swung the machine around and headed for solid ground. But it was too late. Although the tracks were whirling rapidly, they could not gain a foothold on the rapidly disintegrating bank. Slowly the great machine slid back until it tottered on the very brink. There was nothing for me to do but leap from the cab. I did so.

And the next moment, while a cry of horror arose from the crowd, the mighty mechanism went coasting backward into the hideous abyss. Never have I seen a piece of heavy excavating machinery travel so fast. Down it went, like one of those boats on the chute-the-chutes at Coney Island, and it ended with a sickening crash in the bed of the stream far, far below.

As you may imagine, I was somewhat appalled by the magnitude of this catastrophe. And, to make matters even worse, the motion-picture cameraman at once set up an agonized wail, informing me that when I whirled the shovel around, in my vain attempt to save it, I had run over his precious camera. The man himself, being a nimble fellow, had succeeded in leaping out of the way. But the camera was completely smashed and all the film was ruined. This made the disaster complete. There was no picture, and there was no hope anymore of making one. I might have procured another camera, but the only available shovel was lost in the abyss. It was almost more than I could bear, especially when you consider that all the time I had to listen to a continuous yapping from the cameraman.

I tried to explain that the accident had been unavoidable, but the fellow was most unreasonable, claiming that it was all my fault, and that I would have to make good his loss to the extent of something over five hundred dollars. He said that he could show me receipted bills to prove that his equipment cost that much. And he announced that if he did not get full payment he would sue me.

At this point Mr. Blake, the owner of the shovel, entered the discussion, and put on a truly shocking display of bad temper and worse manners. He informed me, in the most insulting way imaginable, that I would have to pay him fifteen thousand dollars for the destruction of his shovel. And when I pointed out that we did not yet know how badly the shovel was damaged, and that it might be possible to salvage it, he came back with a proposal that was even worse.

"All right," he said. "If you will put the shovel together again, and get it up out of that canyon, and pay me for the loss of time on this job, I will let it go at that. But you've got to make this thing right, or I'll sue you."

By this time I realized that I was not getting anywhere with all this talk, so I skillfully closed the discussion by simply turning my back and walking off in a calm and dignified manner. Without so much as turning my head to look back, I made my way to the vacation camp a half a mile distant, and hired a chauffeur and a car to transport the actor, the actress, the male and female clothing dummies and myself back to Los Angeles. Before we left, I noticed Mr. Blake starting out in his car with the photographer, but I paid no attention to them at all.

He informed me, in the most insulting way imaginable, that I would have to pay him fifteen thousand dollars for the destruction of his shovel.

The ride to Los Angeles was uneventful, but when we arrived at the hotel there occurred, as a climax to this day of unprecedented disaster, a final unfortunate episode. As I alighted from the car and thrust my hand into my pocket, I discovered that I did not have enough ready cash to pay either the chauffeur, or the actor and the actress. And at once all three of them began making nasty remarks about suing me. Fortunately, the hotel manager was kind enough to cash a check for me, so I was able to pay them off, and the matter was adjusted. But the whole thing was almost more than my bruised emotions could stand, and I retired to my room in a very low state of mind.

It is now evening, and I have been somewhat revived by a good supper. But I still don't feel any too good. I have abandoned all hope of making a shovel picture. And, as I said at the beginning of this letter, I would like nothing better than to follow your instructions and come hastening back to Illinois. But my high sense of honor tells me that it is my duty to remain here to fight this preposterous claim for fifteen thousand dollars in damages—to say nothing of the relatively unimportant five-hundred-dollar matter. At the moment, my nerves are so shattered that I am unable to plan out a course of action. I imagine that Mr. Blake and the cameraman will call on me very soon—possibly as early as tomorrow. If so, I can only hope and pray that by that time I may have sufficiently repaired my broken morale, so that I can successfully defend myself against what will probably be a very vigorous attack.

I will let you know what luck I have—if any.

Yours, in disconsolate sadness,

ALEXANDER BOTTS.

————

HOLLYWOOD PLAZA HOTEL
HOLLYWOOD, CALIFORNIA

THURSDAY EVENING, OCTOBER 11, 1934.

MR. GILBERT HENDERSON,
PRESIDENT,
EARTHWORM TRACTOR COMPANY,
EARTHWORM CITY, ILLINOIS.
AIR MAIL.

DEAR HENDERSON: It gives me the greatest pleasure to report that all of the fears and doubts which I expressed in my last letter to you have turned out to be absolutely without foundation. All my difficulties have been satisfactorily adjusted. And I find myself once more floating along on the crest of the wave. I am doing so well that I can hardly believe it myself.

This tremendous change in my fortunes began this morning when I received a call from Mr. Blake, the owner of the shovel. When he arrived, I will have to admit that I was prepared for the worst—in fact, for two days and three nights I had been skulking about the hotel in what I can only describe as an exceedingly apprehensive state of mind. You can imagine my surprise, therefore, when Mr. Blake greeted me with the greatest cordiality and started in to thank me for everything I had done for him.

"It seems incredible," he explained, "but that old shovel was hardly damaged at all. It stayed right side up all the way down; it had the good luck to miss all the rocks and big trees; and, although it must have taken a good healthy bump when it ended up in the stream, it is so strongly built that nothing of any importance got smashed. My operator has started it up and it is running great. So I have come to apologize for my hasty words the other day, and I want to thank you for sending the machine down there."

"But why should you thank me?" I asked, somewhat puzzled.

"Because you have saved me so much time in the construction of the dam. Now that the shovel is down at the dam site, I can start excavating at once. I don't have to wait a month or more to build a road into the canyon. Of course, I will need the road eventually, to transport my materials and equipment when I get ready to begin pouring the concrete. So I have just ordered, through your dealer, another shovel to use in building the road."

"Sir," I said, "you surprise and delight me."

"I am surprised and delighted, myself," he said, "at the way this thing is working out. You have no idea how much you have helped me by placing my shovel at the exact point where I wanted it."

I was so knocked into a heap by Mr. Blake's remarks that I could not utter a word. And, while I was still in a considerably dazed condition, the telephone rang.

It was my old friend Mr. Sam Blatz, of Zadok Pictures, Inc. He said he wanted to see me. I pulled myself together as best I could, tottered out of the hotel and took a taxi to the Zadok lot.

Here I received another astonishing piece of news. It appears that the Zadok people, in addition to their regular features, produce a newsreel. One of their news photographers had been sent out to take some movies of the tennis matches which were being held last Monday at the vacation camp near the scene of my shovel exploit. During a lull in the play, he had heard that some sort of excitement was going on at the canyon. He decided to go over, and arrived just in time to get a complete motion-picture record of the sensational descent of the shovel into the canyon. Of course, I knew nothing about this at the time—having been too excited to notice anything—and when I heard about it, it was all a glad happy surprise.

It appears that my flying leap out of the cab took place on the side away from the camera, so it did not show at all, and the picture created the impression that I had driven the machine down that awful slope just as an ordinary part of the day's work. And to make the incident even more striking, the photographer had later climbed down into the canyon and taken a few shots of the shovel when it was started up and tested out by Mr. Blake's operator.

The whole thing made a very nice little newsreel bit. It had already been sent out all over the country. And most important of all, it had been seen by Mr. Blatz, and had converted him to the idea that power shovels might be motion-picture material.

All of this information was given me by Mr. Blatz himself, as a preliminary to a lot of questions about the cost of our shovels, the exact nature of the work which they could do, and their ability to perform other thrilling and spectacular stunts. Mr. Blatz was so interested, so enthusiastic and so absorbed in these technical questions that it was only with the greatest difficulty that I was able to direct his attention to my carefully written power-shovel scenario, which I had brought along, and which I kept thrusting continuously under his nose. At last, however, he took the manuscript and stuck it into a drawer in his desk. So, you see, he has probably accepted it, which naturally pleases me very much.

Taking it all in all, my visit with this great motion-picture magnate was delightfully satisfactory, though Mr. Blatz did not place a definite order for the purchase of any shovels.

Naturally, I came away from the interview in a state of mind which can only be described as sheer ecstasy. And when I got back to the hotel, I felt so encouraged, so delighted, so pleased with myself and so filled with enthusiasm, that I started in immediately to work out a plan of campaign by which I hope and expect to go on to future triumphs which will, I believe, be destined to eclipse even the very real and substantial successes which have already resulted from my efforts.

My new plans are simple, but magnificent. Now that I have succeeded in establishing tractors and power shovels as fit subjects for motion-picture exploitation, there is no reason in the world why I cannot do the same thing with all the other types of machinery which are manufactured and distributed by the Earthworm Tractor Company. I am, therefore, starting in at once to write up a whole flock of motion-picture scenarios, each one of which will present a highly dramatic story based upon the lives of the people who are engaged in handling these machines.

I have already worked out, in the short space of a few hours, the outlines for no less than three complete pictures along these lines. I have one stark and tragic drama dealing with life among the Eskimos, and featuring our entire line of snowplows—straight blade, locomotive type and rotary. I also have a charming pastoral comedy, full of whimsicalities and delightful lyrical passages, dealing with a country lad who wins the hand of his sweetheart by using one of our combined harvesters to cut the grain belonging to her father. In this particular drama I am depending more on atmosphere than on action, but I am using one very exciting incident at the end where the villain is fed through the threshing machine and comes out in the form of chaff which the wind driveth away. I also have a pile-driver story, which is a real humdinger, and which ends with the villain being pounded down into the mud of the river bottom just like a wooden pile. These, of course, are just a few of the ideas which I am going to use. I have many more. And before I get through, I may possibly put across as many as a dozen pictures; thus starting an entirely new cycle.

As you know, motion pictures are often developed in that way. Years ago we had a cycle of pictures dealing with the sheiks of the desert. More recently we have had costume-drama cycles, the gangster cycle and at present we are in the midst of the Federal-detective cycle. My contribution will be the heavy-machinery cycle—pictures dealing with tractors, shovels, combined harvesters, snowplows, pile drivers, graders, dump wagons and so on.

You see, I am really making the fur fly out here. And it is a great satisfaction to me to discover how well adapted I am to the motion-picture business. As a matter of fact, I have no doubt at all but that Mr. Blatz would be only too glad to offer me a permanent position in his organization at a truly princely salary. But as long as I am under contract to act as sales manager of the Earthworm Tractor Company, it would, of course, be unethical for me to leave you people in the lurch. I will, therefore, continue to work along on the same lines as I have in the past.

And I am laying my plans accordingly. As it will probably take me several months to complete work in promoting this new cycle of motion pictures, I am going to look around for a good apartment, and I shall probably send for my wife and the twins to come out and join me.

Yours, with the greatest enthusiasm,
ALEXANDER BOTTS.

————

TELEGRAM
EARTHWORM CITY ILL OCT 15 1934
MR ALEXANDER BOTTS
HOLLYWOOD PLAZA HOTEL
HOLLYWOOD CAL

YOUR LETTER IS RECEIVED AND I AM FORCED TO ADMIT THAT I WAS ABSOLUTELY WRONG AND YOU

WERE ABSOLUTELY RIGHT REGARDING PUBLICITY
VALUE OF MOTION PICTURES STOP APPARENTLY THE
SHOVEL NEWSREEL HAS BEEN RUSHED OUT ALL OVER
THE COUNTRY AND WE HAVE ALREADY RECEIVED
TELEGRAMS FROM SIX DEALERS STATING THAT THIS
IS WONDERFUL PUBLICITY AND THAT PROSPECTIVE
PURCHASERS HAVE BEEN SO IMPRESSED WITH THE
STURDINESS OF OUR SHOVELS AS DEMONSTRATED BY
THE PICTURE THAT AT LEAST FOUR SALES HAVE BEEN
PUT OVER ON THIS ACCOUNT STOP IN VIEW OF THE FACT
THAT YOUR MOTION PICTURE ACTIVITIES ARE DOING
US SO MUCH GOOD AND AS LONG AS YOU WISH TO
CONTINUE THIS WORK WE WILL BE GLAD TO RELEASE
YOU FROM YOUR CONTRACT WITH US SO THAT YOU CAN
ACCEPT THE MOTION PICTURE JOB WITH THE PRINCELY
SALARY WHICH YOU MENTION IN YOUR LETTER STOP
PLEASE WIRE CONFIRMATION SO THAT WE MAY GO
AHEAD AND APPOINT NEW SALES MANAGER STOP
CONGRATULATIONS AND MORE POWER TO YOU

GILBERT HENDERSON

TELEGRAM

HOLLYWOOD CAL OCT 15 1934
MR GILBERT HENDERSON PRESIDENT
EARTHWORM TRACTOR COMPANY
EARTHWORM CITY ILL

HOLD EVERYTHING STOP YOU MUST BE CRAZY IF YOU
THINK YOU CAN GET BY WITH CANCELING MY CONTRACT
STOP I WONT STAND FOR IT STOP HOLLYWOOD IS
WONDERFUL AND THE MOTION PICTURE BUSINESS
IS DELIGHTFUL BUT I AM A NATURAL BORN TRACTOR
MAN AND COULD NOT EVEN CONSIDER A PERMANENT
JOB IN ANY OTHER BUSINESS STOP I AM ARRANGING
MATTERS SO THAT I CAN CONDUCT ALL MY FUTURE

BUSINESS WITH THE MOVIES BY MAIL AND TELEGRAM AND I AM LEAVING BY PLANE TONIGHT STOP WILL BE AT MY OFFICE IN EARTHWORM CITY TOMORROW AND IF I FIND YOU HAVE APPOINTED ANOTHER SALES MANAGER I WILL SUE YOU

ALEXANDER BOTTS

ABOUT WILLIAM HAZLETT UPSON

WILLIAM HAZLETT UPSON was born in Glen Ridge, New Jersey, September 26, 1891. His father was a Wall Street lawyer, his mother a doctor of medicine, and most of the rest of the family doctors, lawyers, college professors, and engineers. To be different, Bill Upson became a farmer—but found the job involved too much hard work. He escaped from the farm by enlisting in the field artillery in World War I.

After the war, he worked from 1919 to 1924 as a service mechanic and troubleshooter for the Caterpillar Tractor Company. "My main job," he said, "was to travel around the country trying to make the tractors do what the salesman had said they would. In this way I learned more about salesmen than they know about themselves." He also became very fond of salesmen. He admitted they are crazy, but maintained they are splendid people with delightful personalities.

In 1923, Bill began writing short stories. In 1927, he created the character Alexander Botts, who has appeared in over a hundred *Saturday Evening Post* stories.

Bill was married to Marjory Alexander Wright. For many years, their home was in Middlebury, Vermont. They had a son, Job Wright Upson, a daughter, Polly (Mrs. Claude A. Brown), and a dog, Shelley. Upson spent his last days in Middlebury with his family, and passed away on February 5, 1975, at the age of 83.

ABOUT ALEXANDER BOTTS*

Requests for biographical material have come from Botts fans, particularly the younger ones who missed many of the earlier stories. From the files of the EARTHWORM CITY IRREGULARS—the national Botts fan club— come the following notes:

ALEXANDER BOTTS was born in Smedleytown, Iowa, on March 15, 1892, the son of a prosperous farmer. He finished high school there, then embarked on a series of jobs—none of them quite worthy of his mettle. In these early days the largest piece of machinery he sold was the Excelsior Peerless Self-Adjusting Automatic Safety Razor Blade Sharpener. He became interested in heavy machinery in 1918 while serving in France as a cook with the motorized field artillery. In March 1920, he was hired as a salesman by the Farmers' Friend Tractor Company, which later became the Earthworm Tractor Company.

On April 12, 1926, he met Miss Mildred Deane, the attractive daughter of an Earthworm dealer in Mercedillo, California. Seven days later they were married. Mildred, later nicknamed Gadget, had attended the language schools at Middlebury College (Vermont) and acted as interpreter for her husband when he was sent to Europe in 1928 to open new tractor outlets there.

Mr. and Mrs. Botts returned from Europe in early 1929 to await the birth of Alexander Botts Jr., who arrived in February along with a twin sister, Little Gadget. Mr. Botts now has been a grandfather for some years. The adventures of Botts have been appearing in *The Saturday Evening Post* since 1927, over a hundred stories in all. They have been collected into eight books, most of them now out of print. Your local bookstore probably can locate used copies of the books, or, if not, the EARTHWORM CITY IRREGULARS office usually has copies on hand. Write to R. D. Blair, 38 Main Street, Middlebury, Vermont, for information, and—if you're a true-blue Botts fan—ask about joining the IRREGULARS.

*This biography was written by William Hazlett Upson and published in the 1963 book Original Letters of Alexander Botts by Vermont Books: Publishers in Middlebury. While the EARTHWORM CITY IRREGULARS are no longer meeting, Upson's local bookstore— The Vermont Book Shop—which opened in 1949 is still open at this address today.

We are pleased to have presented this collection, which is the third installment of the Alexander Botts stories, along with the illustrations that accompanied them in *The Saturday Evening Post*.

We intend to present the full collection of over 100 short stories, including some that were published outside of the *Post*. To see where Bott's next sales call takes him, visit octanepress.com or anywhere books are sold.

Follow Octane Press on Facebook, Instagram, Twitter, or the latest, new-fangled social media platform to learn more!

Manufactured by Amazon.ca
Bolton, ON

26657050R00159